NSF MOSAIC READER

EVOLUTION
NEW PERSPECTIVES

AVERY PUBLISHING GROUP INC.
Wayne, New Jersey

ALLEN COUNTY PUBLIC LIBRARY
FORT WAYNE, INDIANA

The articles contained in this volume were selected from original works that appeared in *Mosaic* magazine. They are reprinted by permission of the National Science Foundation.

Mosaic is published six times yearly as a source of information for the scientific and educational communities served by the National Science Foundation. For more information regarding *Mosaic,* please direct your inquiries to: Editor, *Mosaic,* National Science Foundation, Washington, D.C. 20550.

The publisher is indebted to Warren Kornberg, Editor of *Mosaic,* for his editorial guidance, his invaluable suggestions, and his patience. Avery also wishes to thank the members of its own editorial board for their help in article selection. Our thanks go to Paul Biersuck, Department of Biology, Nassau Community College; Mel Gorelick, Department of Biological Sciences, Queensborough Community College; John Burkart and Loretta Chiarenza, Department of Biology, State University of New York at Farmingdale; Bernard Tunik, Department of Biology, State University of New York at Stony Brook; John Maiello, Department of Biology, Rutgers—The State University of New Jersey; and Donald Wetherell, Biological Science Group, University of Connecticut at Storrs.

Cover design by Martin Hochberg.
Cover photo credit: Martin Hochberg.
In-house editor: Joanne Abrams.

Copyright © 1983 by Avery Publishing Group, Inc.

ISBN 0-89529-175-4

All rights reserved. No part of this publication may be reproduced, stored in a retrieval system, or transmitted in any form or by any means, electronic, mechanical, photocopying, recording or otherwise, without the prior written permission of the copyright owner.

Printed in the United States of America

10 9 8 7 6 5 4 3 2 1

CONTENTS

7055966

Introduction iv

The Evolution of Evolution 1

    Until recently, neo-Darwinism—a synthesis of Mendelian genetics and Darwinian natural selection—was widely accepted as the universal theory of evolution. Although natural selection still enjoys unqualified support, the traditional concept of gradual evolutionary change is now being challenged by that of punctuated equilibrium: a belief in periods of evolutionary stability, interrupted by spurts of speciation and mutation. The author discusses these two schools of thought, examines the evidence cited, and looks ahead to a decade of intensive research and debate.

Life in the Precambrian 10

    Not very long ago, life on earth could not be traced beyond some 600 million years ago: the dawn of the Cambrian age. There must have been precursors, but they seem to have left no trace. Now, researchers have uncovered that missing evidence: bacterial and algal life forms that appear to have existed as far back as 3.5 billion years ago. As new evidence continues to prompt a flurry of investigations, controversies centering on the origins of life become increasingly heated. This reading examines the wide divergence of opinion and interpretation regarding life's beginnings, and discusses how far we've come—and how far we may have yet to go—in our search for the earth's oldest organism.

Molecular Evolution: A Quantifiable Contribution 22

    Traditionally, anthropologists, paleontologists, and biologists have sought to establish evolutionary relationships through two disciplines: comparative morphology and fossil dating. Now a new branch of study dealing with DNA and proteins, not bones and rocks, is providing scientists with a third—and impressively accurate—source of information. The author details the developing methodology of molecular archaeology and explains the assistance it offers researchers working toward a more complete picture of the tree of life.

The World's Great Dyings 32

    The history of life on earth is marked by many of what paleontologists call the great dyings—rapid disappearances of great numbers of the earth's species. The traditional, uniformitarian view that the earth's history is an unbroken sequence of gradual changes cannot sufficiently explain these events. Hence, there has been a resurgence of what is called catastrophism—the willingness to accept at least intermittent catastrophic events in the earth's uniformitarian history, and thereby explain some mass biological extinctions. This article surveys the various theories that scientists have offered to explain the great dyings, and details a new finding that is changing the course of the long-standing debate.

"Feedback" Produces a Theory of Ecology 41

    This discussion centers on coevolution, a concept that is having a marked impact on the direction of evolutionary research. Rather than viewing organisms as isolated entities, coevolution emphasizes the reciprocal evolutionary changes that occur in two or more species as a result of their method of interaction, be it symbiotic, predatory, or competitive. The author examines the intricate interplay of species and environment, citing a number of cases in which plants and animals have impelled each other to change and adapt.

The Significance of Flightless Birds 52

    This reading examines the efforts that are being made to establish evolutionary relationships among related species, using the flightless birds, known as ratites, as an example. These birds, which include the ostrich of Africa, have long constituted an enigma in the scientific world. Researchers have debated the evolutionary processes that shaped the order, have pondered the birds' relative positions on the evolutionary tree, and have even questioned their common ancestry. As a testing ground for the science of molecular species classification, ratites are providing answers about themselves, about the discipline that seeks to analyze them, and about what they may represent.

Glossary 58
Index 60

INTRODUCTION

Mosaic, the source of the articles in this reader, is the bimonthly magazine of the National Science Foundation. Its purpose is to keep nonspecialists in any of the sciences aware of the ferment at the frontiers of many scientific research disciplines and the research trends out of which tomorrow's scientific and engineering advances will emerge.

Mosaic's purpose is to explore the thinking of researchers about both the current and future status of their science. Its articles provide insight into the problems facing investigators in virtually every research area, and explore the ways they seek to overcome those problems—often by crossing traditional disciplinary lines.

Prepared by experienced science journalists and authenticated by scientists, these articles reveal the processes of science, as well as its progress. They not only report on day-to-day advances but offer perspectives on what science is.

For the *Mosaic Reader Series,* recent issues of the magazine have been surveyed, and groups of articles assembled to provide broad, pertinent overviews of segments of scientific research. This reader on evolution as a scientific discipline considers the changes evolutionary theory itself has undergone and is undergoing as research insights and techniques improve. Intact, of course, are the basic tenets of natural selection. But the ways it might express itself and the ways its evidence might be read are both undergoing change as biologists further hone the research methods that serve as their source of information and illumination.

The Evolution of Evolution

by Boyce Rensberger

After 120 years of microevolution, is evolutionary theory poised for a quantum leap? Or does evolutionary theory allow quantum leaps?

In most branches of the diverse field called evolutionary biology, these are times of unusual ferment. To some investigators, the 1980s are likely to be the most productive and exciting years since Charles Darwin's epochal work more than 120 years ago. Failing that, the decade will surely be the most contentious.

It is not the reality of evolution that is being contended; evolution is part of the bedrock of contemporary biology. But the orthodox views of how it comes about are being severely challenged. Adherents of the traditional, or neo-Darwinian, view—that natural selection causes species to evolve gradually, one from another—must now re-examine ideas that have served for more than a century to explain the earth's panoply of life forms.

The idea of gradualism is at the heart of traditional evolutionary theory. It is also at the heart of assaults on traditional theory, and with good cause; little has ever been found in the fossil record to document gradual change from one species to another.

Ironically, the problem faced by gradualism was recognized more than a century ago. Darwin himself conceded that the fossil record did not show much evidence of gradual change. But he was so sure that further discoveries would fill in the transitional forms that he went ahead to propose his theory of natural selection as an explanation of the transitions.

Most species, as Darwin said, tend to produce more offspring than can survive. Thus there is competition for resources. Since siblings vary, individuals possessing traits that give them a competitive edge are more likely to survive and pass on their special traits. Thus a feature that may have appeared as a random oddity in a litter or brood may come, after many generations, to predominate in the species. Over millions of years, Darwin suggested, many small changes would gradually accumulate to produce a descendent species very different from its ancestor.

Darwin knew, of course, about so-called artificial selection, by which plant and animal breeders permit only chosen individuals to reproduce, thereby creating strains in which once-rare features become standard. He reasoned that because nature exposes every new individual to dangers and competitions, there is what one might call a natural selection that favors survival of those best adapted to the environment.

Through the late 19th and early 20th centuries, paleontologists scoured the globe, turning up fossil evidence of thousands of extinct life forms. Some appeared to be intermediates in a series, but many were simply different from anything else. Some of the early evolutionists did claim to perceive long-term trends among those different forms.

The modern synthesis

In 1900, Gregor Mendel's pioneering genetics research was rediscovered after 40 years of obscurity. Mendel had found in his study of generations of pea plants that inherited traits were passed on as discrete units, now called genes. Darwin apparently never knew of Mendel's work, which was advanced by others through elaborate experiments with fruit flies in succeeding decades. By 1930, Mendelian genetics was integrated with Darwinian natural selection, and neo-Darwinism, or the modern synthesis (the current orthodoxy) was born.

The modern synthesis is not a single formula or statement but a broad and variously interpreted set of propositions to which many great scientists have contributed. Among its chief founders were George Gaylord Simpson, the paleontologist, and Theodosius Dobzhansky, the geneticist. Other major contributors included R. A. Fisher, J. B. S. Haldane, Ernst Mayr, G. Ledyard Stebbins, and Sewall Wright.

In its early days, according to Harvard paleontologist Stephen Jay Gould, a principal critic of the idea who is also a historian of science, the synthesis was fairly pluralistic. It was open to a variety of possible modes and tempos of evolution. But by the late 1940s and early 1950s, Gould says, the perspective had narrowed.

As Gould characterizes the modern synthesis—and here he has been accused of simply setting up straw men—orthodox evolutionists during the 1950s and 1960s insisted that all important evolutionary change had been the product of gradual accumulations of small changes that were favored by natural selection. At their most extreme, suggests Gould, modern synthesists held that every feature to be found in an organism, from its overall size and shape down to the finest details of its microanatomy is adaptive.

For the modern synthesists, Stebbins, of the University of California at Davis, insists that their views were never so rigid or extreme. Rather, an emphasis on gradualism guided by natural selection has always been their thesis. And it is a compelling theory. The living world is indeed staggeringly rich with examples of exquisite adaptation, species after species with curious and wonderful shapes, structures, and colors that fit them closely to some narrow ecological niche. "I well remember," notes Gould, "how the synthetic theory beguiled me with its unifying power when I was a graduate student in the mid-1960s. Since then I've been watching it slowly unravel as a universal description of evolution."

Punctuation: another approach

Actually, Gould has had a more active role in what he calls an unraveling than those mild words imply. In 1972 he and Niles Eldredge of New York's American Museum of Natural History published their theory of what they called punctuated equilibria. They argued that after more than a century of searching for the transitional forms Darwin spoke of, a much fuller fossil record showed relatively little gradual change. In many lineages, the record shows no change at all. What the fossils do reflect, Gould and Eldredge argued, is that in most cases species simply appear at a given time, persist relatively unchanged for a few million years, and then disappear. In the meantime, other, similar species may arise and go on to survive longer, but they too usually appear in the record full-blown and then remain morphologically static.

Gould and Eldredge also argue that perceived evolutionary trends, such as the increasing size of horses, were not the result of gradually changing morphology within a species. Instead, they hold, the units being selected were entire species. In other words, evolutionary processes were rapidly spawning morphologically static species, each of them subject—as a whole—to survival or extinction.

Such a contention flatly contradicts a basic tenet of evolutionary theory: that evolutionary pressures select individuals, not species. The synthesists are further outraged by the Gould-Eldredge contention that natural selection itself is a mere epiphenomenon,

Hawaiian fly. *Drosophila* flies serve in the study of evolution. Whether species develop gradually or in spurts is at the center of the current debate.

Hampton Carson.

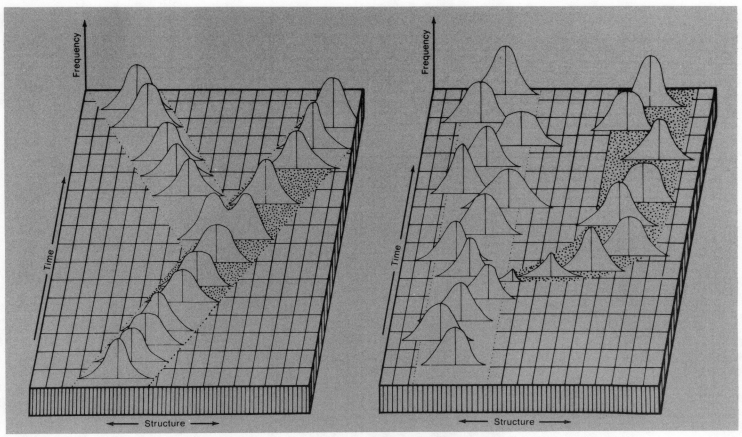

Alternative models. Only morphological structure is considered in these two models of evolutionary change: gradualism (left) vs. punctuated equilibria. The gradualism theory holds that the rate and direction of change is uniformly low and unidirectional, while punctuated equilibria contends change is episodic, oscillating in direction.

E. S. Vrba/ *South African Journal of Science.*

yielding only microevolutionary changes within a species.

Gould and Eldredge declared their new construction of evolutionary theory to be a challenge, an alternative to neo-Darwinian gradualism. They threw down a gauntlet that the synthesists did not take up until the pair restated their case, even more emphatically, in 1977. Microevolution and its obverse, stasis (periods of evolutionary stability between macroevolutionary events) helped spark the arguments that followed.

Microevolution and stasis

A classic example of microevolution involves the peppered moth of England. In the 19th century, the moths' wings were light-colored but peppered with tiny dark patches. They spent much of their time on similarly colored tree lichens, and their coloration was effective camouflage against predatory birds. These light-colored moths occasionally spawned a dark one. Since these were conspicuous to birds, they were likely

---

Rensberger recently joined Science 81 *as a senior editor.*

to be eaten before they could reproduce. But when industrial pollution killed the tree lichens and blackened the tree bark, it was the black moths whose coloring favored them. White moths, picked out and eaten by the birds, became rare.

This transition took 50 years and became a textbook example of gradual natural selection documented in the wild. Followers of the modern synthesis asserted that major evolutionary changes, such as the origin of species—today called macroevolution—were simply the result of many microevolutionary steps like that undergone by the peppered moth.

Gould and Eldredge reject the smooth extrapolation of microevolution to account for macroevolution. Some other, much faster mechanism, they declare, must come into play to punctuate the periods of stasis with macroevolutionary episodes. Not everyone agrees.

Thomas J. M. Schopf, a paleontologist at the University of Chicago, thinks that just because there is little or no change in the bones or shells of a species over time, this does not mean that the entire animal remained unchanged. Soft parts don't fossilize; yet, as Schopf notes, they may change radically while the hard parts remain static. He contends that some lineages may become static. But gradual, continuing change is more common.

Much evidence for stasis is based on double-shelled (bivalve) mollusks, Schopf observes. And he cites a case in which certain extinct mollusks, known only from their shells, had been classified as being all of the same species, until living specimens were found. They turned out to be very different internally. Schopf also cites studies of radiation-induced mutations in mice: Only 10 percent of the genetic changes affected the skeleton. Bones, Schopf suggests, are a poor indicator of the total amount of genetic change a species may be undergoing.

Nevertheless, Schopf declares, there are several lineages in which fossilized hard parts show clear evidence of gradual change. An example is the genus *Kosmoceras*, a nautilus-like cephalopod of the middle Jurassic period. Over millions of years, it underwent steady changes in the size and shape of its living chambers and in its ornamentation. The differences have been resolved into six species within the genus.

Further, "Just because something appears suddenly in the fossil record," says geneticist Russell Lande, also at the University of Chicago, "that doesn't mean it evolved suddenly." Fossil sites are the rare, widely scattered places where conditions are right for preserving bones, he notes, and where erosion has reexposed the fossils. The absence of intermediate fossils, Lande says, could well be explained if the transition took place in some small, geographically isolated region—as partisans of both schools of evolutionary theory often suggest. The animal's sudden appearance might simply reflect its migration into the region being sampled.

Limbs in transition

Lande also cites his own research to counter another argument sometimes made by advocates of punctuated equilibria. They say that certain evolutionary transitions must have occurred in very large jumps because intermediate stages would not seem to have been viable—half-formed wings, for instance, or the stumps of vanishing limbs.

Lande's work was on the loss of limbs by various groups whose ancestors were quadrupedal: whales, sea cows, snakes, and others. How, he wondered, could any intermediate form have represented a successful adaptation? Surely, radical anatomical changes such as limb loss must have resulted from unusual macromutations.

Some punctuation advocates have even revived the views of biologist Richard Goldschmidt. Forty years ago, he suggested that evolution produced "hopeful monsters," progeny that differed radically from their parents and which, if able to survive, might become the founders of new species. No one today seems to accept Goldschmidt's notions per se; his mechanisms of change don't track with modern molecular biology. Such mutations do occur, however, but, the University of California's Stebbins argues, such mutations "nearly always" reduce the vitality or reproductive efficiency of an organism.

Russell Lande adduces evidence from both living and extinct species to show that at least limb loss need not have been the result of any macromutation. For example, the pelvic bones of dugongs (close relatives of the sea cow) that lived during the middle Eocene, around 45 million years ago, were well developed. Their hind legs, though small, seem to have been structurally complete and functional. But progressively more recent specimens show several intermediate stages, including some with little more than rudimentary thigh bones. The only evidence of hind legs left in the modern dugong is a pelvic bone with traces of the sockets in which the thigh bones once fit.

A more detailed sequence of limb reduction is known from living lizards such as skinks. These reptiles range from fully quadrupedal, five-toed species to completely limbless, snakelike creatures. It appears, says Lande, that evolutionary limb reduction begins with the loss of toes, one by one, and proceeds up the leg toward the trunk. In fact, there are so many in-between gradations—fewer and fewer toes, then fewer jointed limb segments, then more rudimentary stumps—that taxonomists have been at a loss as to where to draw the line for purposes of classification. In several cases, an entire range of forms has been classed in a single genus.

Nonetheless, each successive stage in such a gradient is represented by a thriving species. Those with nearly complete limbs

Evolution theorists. Stephen Jay Gould of Harvard argues for punctuated equilibria (top left). G. Ledyard Stebbins, from the University of California at Davis, is a proponent of gradualism guided by natural selection (above). University of Chicago's Thomas Schopf contends that some lineages may become static, but gradual change is more common (left).

walk about normally, while those with reduced or absent limbs tend to have longer bodies and to propel themselves by snakelike undulations. Some limbless species are burrowers. Other lizards may use their half-limbs to raise or anchor their bodies for feeding or mating but then fold them out of the way when moving by undulation.

From such examples Lande concludes that intermediate stages of limb development can be adaptive or, at least, not deleterious. This means that a major evolutionary change such as complete limb loss need not have arisen as a single jump.

Lande has shown that a considerable number of sequential morphological steps can occur in a geologically short time. Under modest selection pressures, he calculates, a lizard lineage might lose one digit every 29,000 years. Even under very weak selection pressure, a lineage might go from complete limbs to stubs in a million years.

Molecules as measures

Nor is morphology the only index of similarities and differences, at least where extant species are concerned. A whole new subdiscipline has been growing up in recent years around one type of evidence for relationships among organisms: the evidence found in their DNA or in the sequences of amino acids in their proteins. The extent to which species show molecular differences is a relatively direct index of the length of time since they diverged from each other. (See "The Significance of Flightless Birds," *Mosaic*, Volume 11, Number 3, and "Molecular Evolution, a Quantifiable Contribution," *Mosaic*, Volume 10, Number 2.)

Using this method, and assuming a constant rate of point mutations along a molecule, Alan C. Wilson of the University of California at Berkeley has produced hypothetical family trees that generally square with traditional phylogenies. Wilson, one of a number of scientists in the field, has also calibrated his phylogenetic trees with absolute dates where a reliably dated fossil record of divergence is available.

The pace of evolution, Wilson has concluded, is slow and gradual for most species, faster and more abrupt for others. "I like the punctuation model," he says, "but I think it doesn't apply very often." A frog lineage, Wilson estimates, speciates once every three million years. Mammals in general speciate five times as fast and primates ten times.

Although Wilson thinks point mutations in such molecules as hemoglobin occur at a constant rate and probably account for the slower rates of evolutionary change, he believes the faster pace among mammals depends on another mechanism. In particular, he suggests, it depends on rearrangements of blocks of gene sequences within chromosomes.

These rearrangements occur in lower species but are seldom passed on; a fairly close chromosomal match is needed to produce viable offspring, and an individual bearing a significant rearrangement would rarely find a mate with a similar sequence. Mammals, Wilson suggests, have smaller populations and social arrangements that encourage inbreeding. Thus the likelihood is greater that they will mate with a closely related individual bearing a similar rearrangement.

Debate on synthesis

Among those who pursue evolutionary studies by more traditional methods, many are convinced by neither the gradual nor the punctuation viewpoint. At the University of Chicago for example, Leigh Van Valen, whose work has ranged from fruit flies to dinosaurs, declares himself "not a gradualist and not a punctuationalist. I don't think the evidence is good enough to resolve the issue." Van Valen believes there is probably a range of evolution rates.

Much the same view is endorsed by two major figures in what is considered the modern synthesis establishment: G. Ledyard Stebbins, a botanical evolutionist, and Francisco Ayala, a geneticist and colleague of Stebbins's at the University of California at Davis.

"We object to Gould because he misrepresents the synthetic theory," Stebbins says. He and Ayala contend that Gould has singled out only the narrowest versions of the modern synthesis (chiefly what Gould considers the later, hard-line positions of Simpson and Dobzhansky) and attacked them as if that is all there ever was to the modern synthesis. "All the concepts espoused by [Gould and other] new paleontologists," Stebbins notes, "are rooted in the synthetic theory." He cites Simpson's early coining of the term "quantum evolution" in the 1930s to support his contention that nothing fundamentally new is being proposed by punctuation advocates. The idea of evolution making big changes in a short time, especially in ecologically unstable environments, has a long history, according to Stebbins and Ayala. They believe that critics of the modern synthesis set up a straw man simply to attract "otherwise undeserved attention."

Further, the two argue, the modern synthesis is compatible with either a gradual or a punctuated pace of evolution, or with both. Stebbins quotes one of his own books, published in 1950, to make his point: "Given selective forces acting at their maximum intensity, a normal rate of mutation and the possibility of occasional hybridization, . . . a new species . . . could evolve in fifty to a hundred generations." By comparison, Gould has allowed that as much as 50,000 years—time enough for 2,500 generations, even in a long-lived species—could be considered a single burst, or punctuation, of evolution.

Such differences in perception of time, many evolutionists observe, account for much confusion. A species that took 50,000 or even 100,000 years to evolve would be considered by a paleontologist to have appeared almost instantly. On the other hand, geneticists who work with fruit flies, turning over many generations a year and producing new species with relative ease, would consider that sort of change as excruciatingly gradual. And, Gould argues, a 50,000-year transition *is* an instant when compared with five million years of stasis; this point has weight among those who believe stasis is the rule in evolution. Like paleontologist Schopf, many scientists in this area think that stasis occurs in some lineages but is not a predominant force.

To support gradualism, Stebbins cites the apparently rapid jump in human brain size: a 500-cubic-centimeter increase from *Homo erectus* to *Homo sapiens*. Taking a human generation as 25 years, he notes, 2,000 generations would have passed in 50,000 years. This could account for gradual increase in brain size, he says, "even if the mean brain size of the population increased by only a quarter of a cubic centimeter per generation."

Nevertheless, Stebbins and Ayala concede, a "shift in emphasis" among modern synthesists may be in order. They agree that macroevolutionary changes cannot be extrapolated or inferred from a knowledge of microevolutionary events. Macroevolution, they say, must be studied and understood on its own terms, not as a prediction from microevolution. Stebbins will even concede that punctuated equilibrium "is the way things are going"—as long, he qualifies, as one puts aside extreme statements.

One place where Stebbins draws the line is on the matter of macromutations, or hopeful monsters. "I want to divorce myself completely from any idea that these changes require mutations with large effects that go on to establish themselves in the species," he declares. While such sudden, radical changes are common, he says, they almost never spread through further generations to become established.

How and why?

Aside from the rate at which new species arise, there is the question of how or why they arise at all. The modern synthesists hold natural selection to be the chief factor: An accumulation of beneficial mutations eventually produces a population so different from its ancestor that its members can no longer interbreed.

Some evolutionary biologists now assert that adaptation may not have much to do with speciation. The events that launch a new species, they believe, are simply events that isolate a subset of a species. Unable to interbreed with the rest of its ancestral line, the offshoot is on its own evolutionary course. Any adaptations acquired through natural selection will remain exclusive to it. Thus it becomes a new species.

Scientists have discovered striking instances in the real world where just such things seem to have happened. For example, some plants have speciated for no apparent reason. Plant geneticist Leslie Gottlieb, of the University of California at Davis, has studied several examples of western North American species whose progenitors are still living. In one case, the flowering annual *Stephanomeria*, the new species is found in only a single locality, where it grows alongside its parent.

Evolutionists. (Clockwise from top left) Leigh Van Valen's research at the University of Chicago ranges from fruit flies to dinosaurs. Francisco Ayala and Leslie Gottlieb study plant genetics at the University of California at Davis. David Wake works with salamanders at Berkeley. Hampton Carson of the University of Hawaii searches for *Drosophila* flies in the Puna Forest.

Wake photograph by Saxon Donnelly.

The two species are nearly indistinguishable in structure and form, and the few novel features of the derivative can be accounted for by a very small number of genetic changes. The new species seems to have been established by a rapid and abrupt process of speciation conveying no readily recognizable survival attribute.

A second example, in the genus *Clarkia*, one of the primrose family, was studied by Harlan Lewis of the University of California at Los Angeles. Lewis found an ancestor-descendant pair that differed in appearance only by a slight change in petal shape, although the chromosomes of the derivative showed gross rearrangement. Lewis was the first botanist to argue convincingly from such rapid speciation in plants that the speciation process need not be coupled to evolution by adaptation.

Gottlieb calls such examples "striking exceptions" to the orthodox view of evolution, whereas Stebbins regards them as another kind of exception. Such an example "is not typical of plant species," he declares. "Most often, related species can easily be told apart and occupy recognizably different habitats."

A chain of niches

At the University of Hawaii, geneticist Hampton Carson has come to similar conclusions after many years of investigating the origin of new species of *Drosophila* flies in the Hawaiian Islands. This is the same genus that includes the fruit flies long used by population geneticists.

Even more than Darwin's Galapagos Islands, the Hawaiian archipelago is a living laboratory for evolution studies. As the most isolated islands on the planet, those islands have hosted relatively few colonizing plants and animals from elsewhere. Consequently, those colonists found so little competition that many of the variant forms born on the islands went on to become species in their own right. This is why Hawaii has so many plants and animals that live nowhere else. For example, among members of the family Drosophilidae, Hawaii has some 800 species, 95 percent of them unique to the islands. In all of North America, by contrast, there are only about 225 *Drosophila* species. Hawaii has nearly a third of the world's 2,500 species of this type. (Ayala uses smaller numbers than Carson does, but the proportions remain the same.)

Why, Carson asked himself years ago, is Hawaii so special? Much of the answer, he found, was in the way the islands were formed. As the Pacific crustal plate moved slowly, it passed over a "hot spot" that intermittently vented massive lava domes. The result is a row of islands, each the top of a volcanic shield and each new island younger than the one before. The oldest of the islands, Kauai, is 5.6 million years old. The youngest, Hawaii, at the other end of the row, first broke the waves a mere 700,000 years ago and is still being enlarged every year or so by the world's two most active volcanos, Mauna Loa and Kilauea. (See "The Earth's Fountains of Fire," *Mosaic*, Volume 11, Number 3.) The major islands are separated by at least 30 kilometers of water, a considerable barrier to migrant terrestrial species.

Through chromosome-mapping and other studies, Carson has worked out the genealogies of a major group of nearly a hundred Hawaiian *Drosophila* species—those informally called the picture-winged flies because of their wing markings. The lines of descent fit beautifully with the sequence of island formation. Kauai, the oldest island, was apparently colonized by species from the mainland or older, distant islands. These could have been blown by winds or arrived as eggs in vegetation stuck to birds' legs. They apparently found little competition and spawned several new species. As each new island formed, it was colonized by arrivals from the previous one. Carson has found that most of the species on any one island are unique to it, suggesting that later colonizing events were rare, but opportunities for speciation were abundant.

Of the 24 species of picture-winged flies on Hawaii, the youngest island, virtually all evolved there some time since its formation 700,000 years ago. Carson believes the founders of each new group of species were single, gravid (pregnant) females blown from the nearest older island, Maui. Representatives of ancestral species on Maui cannot be identified with certainty. But Maui does have flies whose chromosomes suggest a linkage to Hawaiian derivatives—species that apparently came from the same ancestor and replaced it.

Carson concludes that speciation is a phenomenon distinct from adaptation; this idea is part of what is called the founder-effect hypothesis, which involves chance events that isolate one individual (a gravid female) or a very small group from other members of the same species. The founder's original species may have had considerable genetic diversity, but the isolated population will include only an incomplete sample of that diversity. Since the genes of the founder probably had nothing to do with the event which isolated that individual, the fact that those genes would come to dominate the new population must be attributed to chance rather than natural selection.

Early success in the virgin habitat, Carson thinks, probably leads to a population explosion, followed by an inevitable crash as the numbers overshoot the environment's carrying capacity. Laboratory breeding of *Drosophila* has shown that while a single gravid female represents only a fraction of her species' genetic diversity, her offspring eventually will acquire enough diversity to produce several distinct groups of descendents. A population crash might—by chance or by preferentially affecting certain groups—remove intermediate forms, leaving two or three separate new lineages.

A courtship mechanism

Carson has gone still further in identifying precisely the mechanism that might serve to isolate a new lineage. The picture-winged flies, it turns out, have elaborate courtship rituals. Males use various behaviors to get females to accept their advances. Study of the behaviors—inside small glass cages where laboratory-reared males and females are paired and watched—is just beginning, but it is clear that the ritual involves a complex of movements and displays. For example, males circle females. They wave or flutter or buzz their wings. They touch the females' abdomens with their front legs. And there is probably more that human observers miss.

Carson and his colleague at Hawaii, Kenneth Kaneshiro, find that each Hawaiian *Drosophila* species has its own ritual and that females will accept only males that behave according to the species norm. They have also found, however, that males within a species vary in their courtship behavior to the point where some are unable to interest a female. By manipulating the laboratory environments of *Drosophila* populations, Jeffrey Powell at Yale has simulated population crashes and then allowed the survivors to rebuild their numbers. Where one kind of courtship ritual had been the norm, another might be rejected by females of the old population but successful in the new population.

Changes in courtship behavior, Carson and Kaneshiro say, may reproductively isolate groups that otherwise would be members of the same species. Carson suspects that, in the wild, population crashes may split a species into two or more reproductively isolated groups, each of which might become the founding nucleus of a new species.

Carson's findings are often cited in support of the theory of punctuated equilibria because they uncouple natural selection from speciation. But he objects to the idea of evolution making sudden, large jumps. "There's just no evidence for any really sudden genetic changes that produce large effects," Carson maintains. "What population genetics shows is that very small genetic changes, occurring in only a few generations, can be enough to split off a new lineage. Then it evolves quite gradually, acquiring its special adaptations. Paleontologists have provided beautiful evidence for the fact of evolution, but I just don't think they follow the literature of population genetics and biology."

Stephen Gould, one of the paleontologists to whom Carson refers, notes that it is a common misunderstanding that his theory of punctuated equilibrium requires "sudden, large jumps" within the brief time scales used by population geneticists. The theory, Gould says, refers to the way species are distributed in the fossil record, where a transition that took thousands of generations may appear instantaneous. The crucial factor, Gould notes, is that once the transition has taken place, the species then remains unchanged for quite some time, even on the long geological time scale.

The developmental view

Representing still another point of view, perhaps less partisan, is David Wake, an evolutionary biologist at the University of California at Berkeley who has strong interests in development. He complains that too many investigators are so wedded to one theory or another that they rely on "theory demand" arguments, ignoring the evidence of nature.

On the one hand, Wake criticizes gradualists who insist that theory demands all

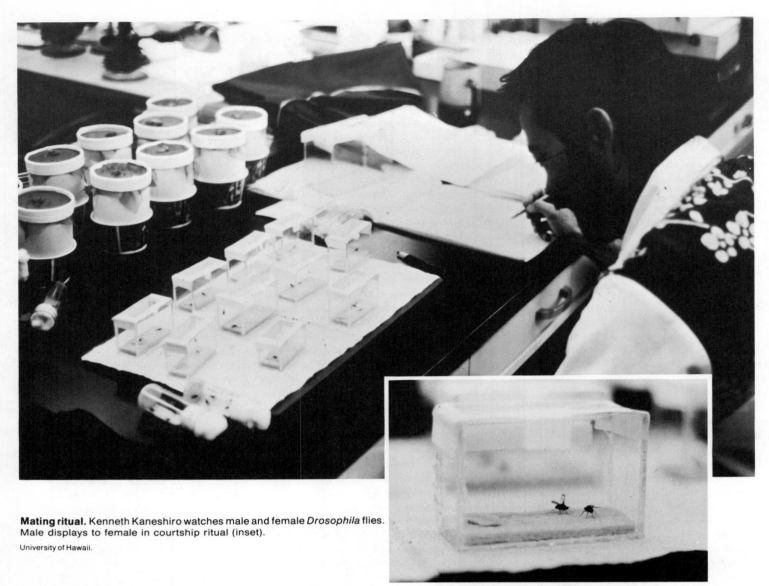

Mating ritual. Kenneth Kaneshiro watches male and female *Drosophila* flies. Male displays to female in courtship ritual (inset).
University of Hawaii.

evolutionary change be achieved in a steady buildup of small increments; he can point to evidence that some morphological changes really do occur in jumps. On the other hand, Wake cites punctuationists whose theory demands that species be so rigidly bounded that the phenomenon of speciation must be the most important event in evolution. Wake does not view species as the most significant entities to study.

"I don't think we know enough yet in evolutionary studies to make rigid theoretical constructs," notes Wake. He agrees that theories are needed to make sense out of what is observed in nature, but he stresses that too many of his colleagues are reluctant to discard or revise their theories when the data do not fit.

Wake himself thinks that the idea of units such as species is overemphasized—that species, as conventionally defined, are not the most significant units in evolution. More important, he feels, are developmental programs: the portion of an individual's genetic endowment that governs its embryonic development, controlling the morphology of the fully developed organism.

As an example, Wake cites two species of salamander descended from a common ancestor. They branched so long ago that they hold only 15 percent of their genes in common; it is genetically impossible for the two species to interbreed. And yet in appearance they are indistinguishable from each other. Of the hundreds of different gross morphologies that characterize salamanders around the world, these two species, although genetically far apart, retained the same set. "In all the evolving that these salamanders have undergone, nothing has touched the genes that affect development," Wake maintains.

Much the same situation exists among many other types of organisms. Such morphologically similar species (called sibling species) would, if found as fossils, be classified together by paleontologists. "It's not true that each species has its own unique morphology. Rather, there are morphologies that transcend species barriers. That's what impresses me," Wake adds.

The insights of developmental biologists like Wake and his student Pere Alberch, now at Harvard, have made a strong impression on scientists in other fields. Many look to this burgeoning discipline to make some of the most important new contributions to understanding evolution. Ledyard Stebbins, defender of a pluralistic modern synthesis, says, for example, "What we need now is a new taxonomy, not of species but of developmental programs."

Few of many

One of the great challenges to developmental biologists is the fact that all the world's biological diversity is confined within only a relatively small number of basic types of organisms. As Harvard geneticist Richard Lewontin puts it, "Most conceivable organisms don't exist." Even if all the known extinct forms are included, it turns out that each is simply a minor variation on one of the known and few basic body plans.

There are, to cite a broad example, no six-legged vertebrates. Or a far narrower example: The cystic arteries that serve the human bladder branch off the hepatic artery to the liver in six architecturally discrete ways. Both these cases, along with hundreds of other examples, Wake argues, represent the final result of developmental programs that can follow only a limited set of pathways.

Consider the cystic artery. All six forms appear to be equally effective in getting blood to the gall bladder. They differ in seemingly trivial ways, such as where the cystic artery attaches itself to the hepatic artery. Yet autopsies of thousands of human cadavers have shown that there is not a gradation of branching points; everybody fits neatly into one group or another.

Why? "I haven't the foggiest idea," Wake acknowledges, "but what this tells me is that the frequencies are much too large for these to be new mutations. I can only assume that these represent slight but discrete differences in human developmental programs. When the cystic artery is being formed, there are only certain ways it can go."

Many other similar examples are known—some seemingly trivial, like the cystic artery, some probably more significant, such as which of a salamander's foot bones fuse. One fusion will produce a foot for grasping, and a climbing animal follows. Other fusion patterns would have other effects. Together, these examples lead Wake to conclude that when an organism's genes begin to express themselves, only certain pathways will be open.

Geologists know that minerals will crystallize only in certain specific ways; they understand the limits in terms of the properties of individual atoms and relatively simple environmental conditions during crystallization. Developmental biologists can only dream about gaining the same level of understanding as to how vastly more complex molecules of DNA map themselves—through a complex embryonic development in a complex environment—into the fully developed, complex organism.

Limited possibilities

Because embryonic development appears to follow a limited repertoire of discrete pathways, Wake says, many different gene combinations can lead to the same morphology: "There are only a limited number of solutions to biological problems." Wake relates this statement to the well-known phenomenon of threshold effects in embryonic development. If, for example, a given trait is influenced by ten genes but has only two morphologies available to it, evolution may proceed apace, altering gene after gene without changing the morphology. Suddenly, after a threshold is crossed, the other developmental pathway will open and, seemingly in one step, there will be a quantum morphological change.

Alberch, Wake's former student, has argued that effects such as this may cause gradual but hidden microevolutionary steps to look like a sudden macroevolutionary jump. As with other bold assertions in current evolutionary biology, this one is both hotly challenged and warmly hailed.

Strong defenders of neo-Darwinism—Chicago's Russell Lande, for example—contend that the appearance of canalization could easily be the product of natural selection rather than any inherent limits to embryological development. "I have a high opinion of what selection can do," Lande says, suggesting that the missing intermediate developmental possibilities may simply have been weeded out as maladaptive.

For a field that only a few years ago was considered by most outsiders and some insiders to be almost cut and dried, evolutionary biology has fairly exploded with new ideas and new evidence and, just as important, with renewed interest in some of its best old ideas and old evidence. "I think the 1980s are going to be the most stimulating decade in evolutionary biology since the 1850s," Wake asserts. "I'm absolutely convinced of it and I'd hate to see us get hung up on trivial debates."

Many researchers would agree about the promise of the coming decade but, of course, those involved in the debates are not likely to see the issues as trivial. William Provine, a Cornell science historian who specializes in evolutionary sciences, believes the current issues are refractory ones. "I don't foresee any quick resolution to the debate," he says. "There doesn't seem to be any one view that slices through the morass and puts everything into perspective." •

The National Science Foundation contributes to the support of research discussed in this article principally through its Systematic Biology Program.

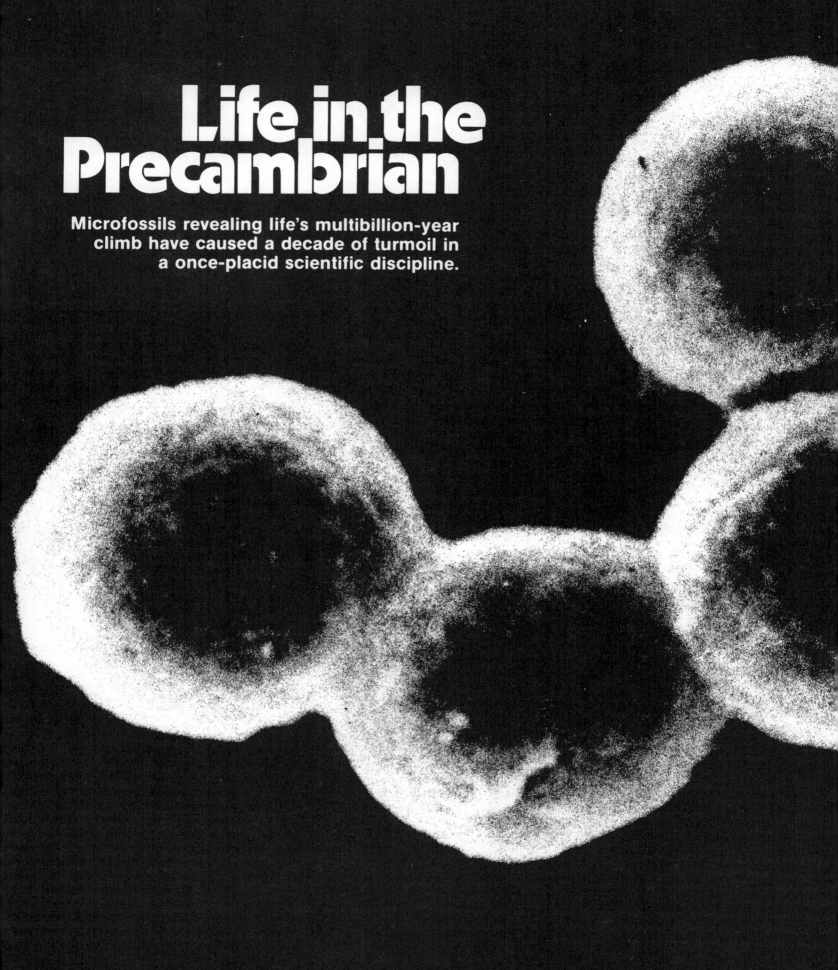

Life in the Precambrian

Microfossils revealing life's multibillion-year climb have caused a decade of turmoil in a once-placid scientific discipline.

Some 3.5 billion years ago, as final drops of lava oozed from the dying moon and the lunar pulse flickered out, the billion-year-old earth was at its morphological best. It was vigorously throwing up mountain ranges, extending continental margins, rounding off ocean basins and filling out its atmosphere. The evidence for this intensive terrestrial activity can be read in those occasional outcrops of 3.3- to 3.5-billion-year-old virgin rock that have been found in Africa, Greenland and North and South America.

Indeed, as scientists study these very old formations, which somehow have remained unchanged by the dynamic processes at work within the earth, they are discovering that still another form of activity was under way then: life. Very simple life, to be sure, but living structures nonetheless.

The precise antiquity of the most ancient of these forms, reminiscent of nothing so much as some primitive bacilli and algae that survive today, is still the subject of heated dispute among paleobiologists. But the fact that there can be a dispute, that physical evidence can be seriously put forward and seriously considered for the presence of life on earth between 2.0 billion and 3.5 billion years ago, itself represents a young and still very active revolution taking place in the science of paleobiology.

Until recently, paleontologists could trace the evolutionary tracks of living organisms back only some 600 million years, to the fossil imprints of such primitive, hard-shelled invertebrates as trilobites. But there the trail went cold. Clearly, the Cambrian trilobites were not the first living things; they were too

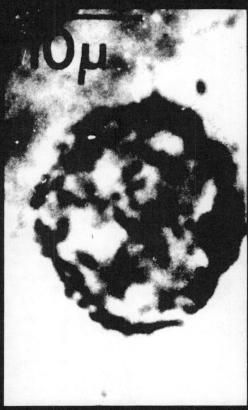

Candidates. Organic spheroids (above), either of biologic or prebiologic origin, are some 3.2 billion years old; the candidacy of these Swaziland, South Africa, samples for election as the earth's earliest formed life is in dispute. The fossils of late Precambrian microflora (left), about 850 million years old, from Australia, are not in dispute.

complex. There had to have been ancestral organisms, each somewhat simpler than its progeny, predating these oldest fossils. But if the immediate ancestors were only a little simpler than those creatures that left imprints in rocks, shouldn't they also have left a fossiliferous mark? If they were, they didn't.

Various explanations were advanced to account for the lack. Some scientists suggested that rock strata older than some 600 million years, in which precursor organisms would have fossilized, had long since been heated and altered by the earth's dynamism; those older fossils would have been destroyed along with the rocks that held them. Others suggested that there had been a long progression of such soft-bodied organisms as jellyfish and worms leading up to the chitinous (hard-shelled) trilobite and its contemporaries; by virtue of their soft structure, they would have been unable to imprint their ghostly shapes in rock layers as they formed.

But from the evidence that is now being accumulated, it appears that something quite different happened. Evolution is being found not to have been a slow, stately parade from its origins two or three billion years ago up a gently inclined ramp of time, with slightly more advanced organisms succeeding each other, small adaptation by small adaptation, at every step of the way. Rather, life appears to have been one sort of thing during much

Event or precondition	Likely stimuli	Probable consequences	Time of event (yrs × 10^9 BP)
12. Tissues, organs	Sexuality, competition, increased O_2	Metazoa, higher plants, changed $CaCO_3$ patterns, land biota, intensified weathering, subsequent evolution	~ 0.7
11. Eucaryotic sexuality (meiosis)	Eucaryotic cell, increased free O_2	Complete eucaryotic hereditary mechanism	> 0.7 probably > 1.3 possibly > 1.5
10. Eucaryotes (mitosis)	> 1% PAL free O_2, shielding of intracellular anaerobic functions	More O_2, CO_3^{--} and SO_4^{--}, less CO_2. Higher algae	< 2 > 1.3 possibly > 1.5
9. Fully oxidative metabolism (superoxide dismutase)	1% PAL free O_2, enzymic neutralization of H_2O_2 and O_2^-	Increased O_2, ozone shield, red beds, eucaryotes	~ 2
8. Cyanophytes (or protocyanophytes)	Heme proteins, Mg porphyrin, biophotolysis of H_2O	Microaerophilism, BIF, CO_3^{--} increase, finally O_2 increase and fully oxidative metabolism	~ 3.8
7. Primitive autotrophs	Competition for components and energy	Cyanophytes, subsequent evolution	
6. Life (anaerobic heterotrophs)	Energy transfer, genetic code	Procaryotic diversity, biogeochemistry	
5. ATP, RNA, DNA		Transient negentropy, reproducibility	
4. Sugar-phosphate bonds, nucleotide bases	Chemical evolution, autocatalysis	ATP, DNA, RNA	> 3.8
3. Amino acids, polypeptides			
2. Simple organic molecules (HCN, HCHO, etc.)	UV irradiation of O_2 free atmosphere	Amino acids, polypeptides	
1. Atmosphere and hydrosphere	Outgassing of earth	Chemical evolution	

Time lines. Way stations on the road to higher organisms as delineated by Preston Cloud (above), and the known geological distribution of Precambrian microfossils (right) according to J. William Schopf.

Call it biogeochemistry

"... Biogeology is the comprehensive term. In addition to structurally preserved fossils and their imprints (paleobiology and paleomicrobiology), its subject matter includes their tracks and burrows (palichnology), their associations in the rock and evidence of their feeding habits and interactions one with another (paleoecology), geochemical evidence both as to the composition of biological materials and as it bears on biological processes in geology, and the broader interactions between evolving biosphere, atmosphere, hydrosphere and lithosphere. It is true that some who conduct their research on a level that is equally or nearly as broad have used the term paleobiology to describe what they do—presumably unconscious of its narrow origins and restricted implications...."

Preston Cloud

of the Precambrian: microscopic forms in which not much change took place over the billions of years and in the existence of whose fossils hardly anyone would believe not much more than a decade ago. And then, dramatically and radically, some 570 million years ago, "macro-life" suddenly took off.

At least in geologic terms, the transition from the bacterial and algal life forms now known to have existed in the Precambrian to the more complex creatures at the beginning of the Cambrian was quite abrupt. It might have spanned no more than 100 million years. So, while the predecessor forms to the trilobites and the other fossilized macro-organisms of the Cambrian explosion may have been soft bodied and therefore unlikely to have left a fossil imprint, it could also have been that their sizes were not as great as traditional paleontologists would have expected to find in a trilobite's ancestor. Nor were their numbers as great as either those that went before or those that came after.

The quest

With the existence of Precambrian microfossils now confirmed, the search for the remains of the earth's oldest living organisms has become something of a quest for a biogeochemical grail. It has led paleobiologists like Elso Barghoorn,

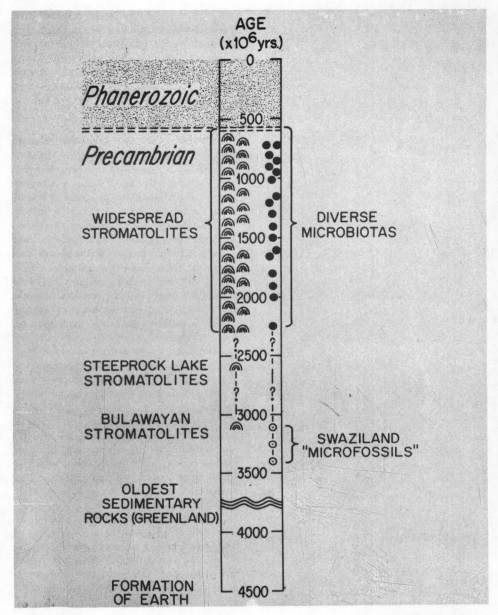

Dawn-Life

Precambrian and Cambrian are geologic divisions of time, the former representing the first 87.5 percent of the earth's history and the latter the beginning of the more recent 12.5 percent. If the earth's 4.5 billion years of age were set equal to one day, the Precambrian would extend from just after midnight, through the morning and afternoon, and into the evening until 9 p.m. It would thus account for 21 of the day's 24 hours. The rest would be the three hours between 9 p.m. and the end of the day: the present.

Preston Cloud, J. William Schopf and others to journey to remote reaches of Africa, Australia, Canada, Greenland, India, the Soviet Union and the United States in search of old but pristine geological formations. For, where the oldest unaltered rocks are, there will be found the fossiliferous traces of the oldest organisms to have inhabited the earth.

The quest has so far produced several fossil grails. They range in age from about 2.0 billion to 3.5 billion years, and each has a devoted corps of discoverers and advocates arguing for its primacy. Barghoorn, Fisher professor of natural history at Harvard University, has found objects which, he maintains, are microscopic fossils at a pair of sites 46 kilometers apart in the Swaziland region of South Africa. The rock formations in which these microfossils were found have been dated as being 3.2 billion to 3.5 billion years old. He firmly believes the imprints in these rocks represent the oldest, most primitive specimens of life yet discovered. His reasoning is sound. But such is the state of today's paleobiology that just as sound appears to be the reasoning of those of his colleagues who fail to share that belief.

Preston Cloud, professor emeritus of geological sciences at the University of California at Santa Barbara, has his own candidates—at least presumptively, on a rating scale for evidence from "permissive" through "presumptive" to "compelling." With that caveat, he prefers cyanophytic (blue-green) organisms responsible for some of the earliest known stromatolitic structures: those found in the lower Pongola system of South Africa's Natal Province. The accepted age for these is about three billion years.

Schopf, a professor of paleobiology and geophysics at the University of California at Los Angeles (UCLA), prefers certain rock structures, presumably also formed by primitive blue-green algae or bacteria, at Steep Rock Lake in Canada and at Bulawayo in Rhodesia, which have been put at between 2.5 billion and 2.7 billion years of age. He, additionally, agrees with Cloud about the candidacy of the Pongola formation, though he has his own probability scale.

A point on which the paleobiologists appear to agree unequivocally, however, is that this is a field very much in ferment at the moment. Of the three dozen or so ancient microbiotas (assemblages of primitive, microscopic life forms) now at least tentatively identified, all but two have been reported since 1965; more than half have been disclosed since 1973, and a third have been known only since 1974.

These findings, many of them by Schopf and his colleagues at UCLA, must, of course, stand the test of independent corroboration; the biological or nonbiological characterization of any microfossil is a delicate business. But because the rocks in which these fossils were imbedded cover the history of the earth between some 570 million years and 3.5 billion years ago, there is confidence that from somewhere among these findings will emerge answers to questions such as:

- From how far back in the earth's history will we find evidence of life, and what clues will that offer to the earlier origins of life?
- What were the organisms like that took part in so incredible a process?
- How did more complex organisms supplant their simpler predecessors?

oxygen and argon? (This question bears upon the issue of when eucaryotes first evolved; some scientists feel that the earth had to have an oxygenic atmosphere before these more advanced life forms could come forward; others maintain that the two probably had to arise in concert.)

A science turns over

Ancient life, or at least the perception of ancient life, was a lot simpler just 25 years ago. "Back in the early nineteen-fifties," says UCLA's Schopf, "the best guess was that life had originated about a billion years ago. And that was just a guess. People thought, 'Well, let's see, if it took 500 to 600 million years to go from trilobites to humans, then it probably also took 500 to 600 million years before that to go from protozoans [single-celled animals] to metazoans [multicelled animals].' Add the two together and you get 1.0 or 1.2 billion years. And that was the figure most people would accept then [in the nineteen-fifties] for the origin of life."

There were suggestions to the contrary. Cloud recalls that stromatolites, even in the fifties, were believed by some to extend back at least two billion years. And as far back as 1922, John W. Gruner had declared that microfossils could be found in Precambrian formations, though he was politely ignored by his colleagues. The conventional wisdom in the fifties still failed to encompass Precambrian microfossils.

But in 1950 a discovery was made that was destined to shatter this consensus. The late Stanley Tyler, a University of Wisconsin geologist, had been exploring old iron ore prospect pits in upper Michigan for a mining firm when he stumbled upon some unusual rock formations. They had a coallike appearance and resembled anthracite coal, which is relatively recent in geological terms. Tyler, for all his knowledge and experience, could not imagine what sort of geologic processes might have produced coal in what were assuredly Precambrian rocks. Biological activity seemed the only reasonable assumption. But these rocks were pretty well dated as being about 1.9 billion years old, and that was almost a billion years more ancient than most paleontologists were then willing to grant for the presence of life.

At a scientific meeting late in 1950, Tyler described the unusual formations to William Shrock, then chairman of the Massachusetts Institute of Technology's geology department. Shrock suggested that Tyler contact a young Harvard botanist, Elso S. Barghoorn.

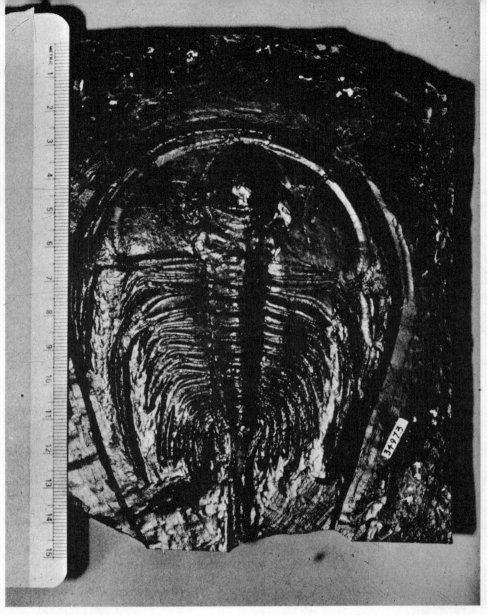

Fossiliferous milestone. Hard-shelled trilobites like this *Fremontia fremonti* from the lower Cambrian were the oldest known fossils until the explosion recently of awareness of Precambrian microfossils.

(Scientists believe that the first living things were heterotrophs, microorganisms biochemically incapable of synthesizing all of the organic compounds they required. Nature probably provided what these simple forms could not synthesize themselves: amino acids, sugars and other organic compounds. These would have been created by the interactions between the primordial atmosphere and hydrosphere [both unlike what they are today] and such energy inputs as solar radiation, lightning and volcanoes. Later, autotrophs arose. These would have been equally tiny organisms, capable of assembling their requisite organic compounds by photosynthesis, using carbon dioxide drawn from the atmosphere. The emergence of autotrophs would have been a major milestone in the evolution of life and, in some views, of the atmosphere too.)

• When did nucleated cells arise, with their genes neatly wrapped inside a second, separate membrane within the cell structure? (Scientists call this cell type eucaryotic, meaning truly nucleated, and its appearance is probably the single most important turning point in the development of life. For, unlike the procaryotic [lacking a nucleus] autotrophs, which evolved first but which had little capacity for evolutionary growth, it was the eucaryotes and their method of sexual reproduction which set in motion the process of evolutionary diversification. It was that process that eventually brought forth trilobites and blue whales, warblers, redwoods and humans.)

• When did the earth sweep out its primordial atmosphere, filled with gases like carbon dioxide and hydrogen sulfide, and begin replacing these with nitrogen,

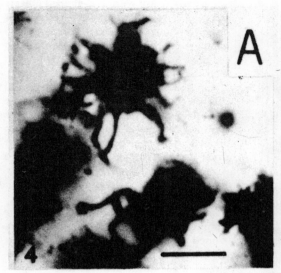

Then and now. Comparison of Gunflint microfossil (left) and living bacterium (center) strengthens the case for the 1.6-billion-year-old sample, and some older. The oldest stromatolite (right), from the Pongola system in South Africa, is 3 to 3.1 billion years old and is indirect evidence of early biospheric evolution.

The two scientists got along well together from the start, Barghoorn recalls. During the next two years they worked on the Precambrian coal, concluding then, though the data were not published until 1957, that it indeed was Precambrian and that indeed it was of biological origin. (The mass necessary for coalification, Barghoorn explains, is assumed to have been "a fortuitous windrowing of algal masses," similar to those that can still be found produced by wind or wave action.)

It was in the late summer of 1952, says Barghoorn (whose correspondence files, since Tyler's death, hold perhaps the only record of these events), just following their field work that year in upper Michigan, that Tyler, working alone, found what has come to be called the Gunflint Iron Formation, on the north shore of Lake Ontario.

Recognizing their potential, Tyler cut wafer-thin slices of the rock, a type known as chert—a compact, sedimentary form of quartz—and in 1953 sent them to Barghoorn. The Harvard biologist put the slices in a microscope and was flabbergasted by what he saw. "I realized," he recalls years later, "that what I was looking at were organisms that looked exactly like fossil algae." Others might have seen the same thing, had they thought to examine Precambrian rocks for signs of microscopic fossils. But so entrenched was the belief that Precambrian fossils would be as visible to the unaided eye as are Cambrian relics, that the evidence of the microscope was either not sought or not credited.

Further testing by Tyler and Barghoorn corroborated the presence of ancient microscopic life forms in the Ontario samples. The two scientists found, for example, that the microfossils contained approximately the same ratio of carbon-12 to carbon-13 as do modern blue-green algae. Living things prefer carbon-12, the lighter isotope of carbon, to the heavier variant, and concentrate more of it in their tissue. Relative to the atmosphere, which contains a great deal of carbon-12 and small amounts of carbon-13, and which is the standard of comparison, living or once-living organisms will possess as much as 25 parts per thousand more of the lighter element than, say, limestone.

A revelation verified

When the results of tests like these became known, and when another widely respected paleontologist, Preston Cloud, independently repeated the Tyler-Barghoorn finding with samples taken from the same locale, resistance to the idea of Precambrian microfossils began to crumble. It became clear that the cherts from this geologic formation, known as Gunflint because early European settlers had found abundant quantities of flint there for use in their firearms, were revolutionary. In Barghoorn's words, they were "a massive accumulation of organic material, of biological material, in rocks known to be Precambrian."

The strange domes were thus revealed to be not so strange after all. They were classified as stromatolites, or algal reefs. Blue-green algae built them, layer by layer, as those tiny organisms of long ago adhered to the surface of boulders in a lake or early sea, the waters of which were saturated with silica. When the silica deposited onto the photosynthetic algae, screening out the sun, others of the plantlike organisms simply climbed up on the silicified backs of previous generations to receive the essential sunlight. And the process lasted as long as conditions were right.

Blue-green algae today still form stromatolites under similar circumstances in some parts of Bermuda, the Bahamas, Florida and Australia, where conditions—principally mineral saturation and the absence of predators that dine on algae—allow. Both ancient and modern stromatolites are remarkably similar; the key difference is that there were more such reefs billions of years ago, before there were snails and other higher life forms around to devour the algal colonies before they could build them.

The upshot of all this was that it imparted to terrestrial life an age greater than most scientists thought possible. Moreover, it touched off a flurry of investigations. If microfossils were to be found in rocks dated at 1.9 billion years, were there any to be found in 2.3-billion-year-old rocks? 3.0 billion-year-old rocks? In addition, and perhaps most importantly, it infused a traditional scientific discipline with enthusiasm that drew to it young, eager scientists and new ideas.

To Fig Tree

One such recruit was J. William Schopf. As an undergraduate majoring in geology at Oberlin College in Ohio during the early nineteen-sixties, Schopf was aware of Tyler's and Barghoorn's preliminary 1954 report on the Gunflint formation. (A more detailed paper on their collaboration

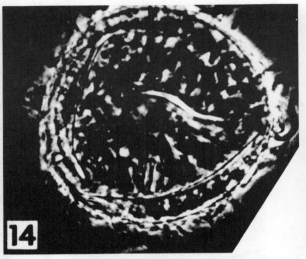
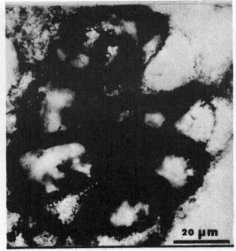

Nucleated organisms. Early eucaryotes, about 500 million years old (top left), were found near Leningrad; large-diameter unicells (right and lower left) come from cherts, 1.3 billion years old, in Beck Spring, California.

did not appear until 1965, a year after Tyler's death.) He had, moreover, become so fascinated with the question of life's early stages, as a result of reading Charles Darwin's *Origins of Species* and George Gaylord Simpson's *The Meaning of Evolution*, that he decided in his sophomore year to become a paleobiologist, a specialist in ancient life forms. He was particularly interested in the Precambrian.

While still at Oberlin, Schopf corresponded with Barghoorn. He entered Harvard as a graduate student in 1964 and, with Barghoorn as his adviser, assisted in Barghoorn's research on another assemblage of Precambrian microfossils.

That assemblage had come from a region in South Africa on the border between the Republic of South Africa and Swaziland. Since the nineteen-fifties, when radioisotope dating techniques were found to be accurate indicators of the ages of rock formations, geologists had been actively searching the earth for its oldest rocks so that they might piece together a better picture of the planet's evolution.

Proceeding on the assumption that the oldest rocks might also harbor the oldest organisms, the paleobiologists—or biogeochemists, as some prefer to call themselves—followed close behind the field geologists. And because radioactive dating techniques had placed an age of 3.2 billion years on that South African border region in the Barberton Mountainland, and because it had been found to be richly carbonaceous, it was there that Barghoorn and others elected in the mid-nineteen-sixties to search for additional Precambrian microfossils. Barghoorn was specifically interested in a series of rock layers exposed by mining operations and called the Fig Tree series, after the name of a nearby brook.

The Fig Tree rocks included several different kinds—chert, ironstone, slate, shale and graywacke (a type of sandstone). Although there were no stromatolites in these layers, some of which were 400 feet thick, Barghoorn nevertheless believed that they once had been covered either by a shallow sea or by a series of marine embayments. More to the point, the carbon-12/carbon-13 ratio of the rocks indicated a biological origin for that element.

In the summer of 1965, Barghoorn brought back to Harvard a number of Fig Tree rocks, thin sections of which were cut and examined under both light and electron microscopes. When he and Schopf studied these slices under the white light microscope, they could make out laminations of organic material, laminations that appeared to have been formed as part of a typical sedimentary process. They could also see spheroidal bodies, 17 to 20 micrometers in diameter, that seemed to resemble cells of microscopic algae.

When the two peered at preparations of the rock under an electron microscope, other specific, more minuscule objects emerged. These were rod-shaped objects ranging between 0.5 and 0.7 micrometer in length and 0.2 and 0.3 micrometer in diameter, objects that in general form seemed to resemble small, rod-shaped (bacillar) bacteria.

On some of the rods, Barghoorn and Schopf could make out walls similar to the cell walls of modern bacteria. From this they concluded that these were more than likely to have been primitive bacteria. The rods were christened *Eobacterium; eo-* is a Greek root for dawn.

"Peas" in "pods"

The two men also thought that the spheroidal bodies, some of which contained blackened interior particles, exhibited many similarities to modern blue-green algae; they named these objects *Archaeosphaeroides barbertonensis*, signifying ancient spheroids from the Barberton Mountainland.

There was, at first, an inclination, according to Barghoorn, to regard the blackened "peas" within the fossilized "pods" as possibly being the decomposed remains of cellular nuclei, a finding that, were it to stand, would throw the field of paleobiology into chaos. Cells with nuclei (eucaryotes) include 99 percent of the organisms extant today; they occupy the top rungs of the evolutionary ladder. And their cytoplasm, the bulk substance of the cell surrounding the nucleus, contains a great many well-defined bodies such as mitochondria and chloroplasts, having highly specific functions within the cell. Indeed, the complexity and sophistication of eucaryotes is clear evidence of the fact that they had to be the result of some highly complicated processes imposed on the earlier, simpler procaryotes.

Procaryotes are very simple forms of life; bacteria and blue-green algae are examples of surviving procaryotic organisms. Such cells have an outer membrane but not an encapsulated nucleus; their genetic content is simpler and usually smaller than that of eucaryotes. Unlike eucaryotes, they usually reproduce asexually simply by dividing their DNA in half and splitting to become two identical daughter cells. Most eucaryotes, on the other hand, reproduce sexually, passing on to progeny genes from each parent. The result is infinitely greater genetic diversity; it is the key to the variety of life that has filled the earth.

It is accepted that procaryotes were the first living organisms on earth and that eucaryotes evolved from them in fairly recent times (possibly as late as 570 million to 850 million years ago). For nucleated cells to have been around more than 3.0 billion years ago would mean either that procaryotic cells would have had to exist long before that—and the earth is only 4.5 or 4.6 billion years old—or that the transition from simple organisms to complex ones occurred soon after life itself got started. If the latter, it almost necessarily would imply that advanced forms of life first appeared around three billion years ago, but then spent two billion years or so in a state of quasi-dormancy before erupting in metazoan profusion in the Cambrian.

The bottom(s) of the pile

Neither explanation would stand up, and Barghoorn and Schopf quickly discarded the idea of eucaryotic fossils in the Fig Tree samples. They reported in 1967 that the blackened fragments were probably "the coagulated, coalified remnants of cellular cytoplasm." Schopf no longer goes along, but Barghoorn remains firmly convinced that these spheroids are the microfossil remains of procaryotes.

Indeed, just last autumn Barghoorn and Andrew H. Knoll of Harvard reported the discovery of a new set of spheroidal microfossils, gathered from a different part of the Fig Tree formation and dated at some 3.4 to 3.5 billion years old. Barghoorn and Knoll are both convinced that these spheroidal objects are fossils either of bacteria or blue-green algae that were trapped and preserved when fine-grained particles settled out of the sea water and entombed them. As they tick off their reasons:

- The organic matter of the sediments, when subjected to chemical assays, yields the carbon-12/carbon-13 ratio typical of modern organisms. Moreover, there are spheroids carrying those blackened "peas" inside their perimeters, which suggests that these were not simply empty bubbles, but were objects containing internal material;
- Their size spread is between 1.0 and 4.0 micrometers, with the mean around 2.5 micrometers. Biologists call this size distribution "a narrowly defined population," which is another way of saying that one would expect the size of similar, primitive organisms to fall within a limited range;
- Their shapes are not perfect spheroids; they exhibit the sort of irregularities one would expect to see in biological matter; and finally,
- Some of the organisms appear to have been engaged in the process of asexual reproduction when they were trapped by the sediments and fossilized; they have the characteristic pinched-at-the-waist shape of simple cell division.

Taken all together, this evidence led Barghoorn to declare: "The only logical explanation is that [the spheroids] are biological." And not just biological from a long, long time ago, but the oldest biological specimens now known to science.

"Fig Tree," he declares, "looks like the bottom of the pile."

And perhaps it is. But not all members of the paleobiological community are willing to accept the Fig Tree as the oldest example of terrestrial life. "The reason that it's so difficult to be sure that these [the Fig Tree findings] really are fossils," observes Schopf, "is because we don't have, as of now, an evolutionary continuum going all the way back to them."

A continuum has been strung, he goes on to say, from the start of the Cambrian, around 570 million years ago, to the Bitter Springs formation in Australia, around 850 million years ago, to the Gunflint deposits, around 1.9 billion years ago. "We've been filling in the gaps between these major findings," Schopf says. "We now have, on the average, one deposit [of

Rhodesian stromatolite. Stromatolitic structure, believed to be of bacterial or algal origin, from 3-billion-year-old rock near Bulawayo.

microfossils] every 50 million years. So we can go from one to another pretty easily. But the Fig Tree is sitting way out there at three billion or more years, and we don't have any intermediate fossiliferous deposits between it and the next oldest site."

In a very gingerly way, Schopf has sought to put some distance between himself and his former adviser, and even the work they once did together. Just a few years ago, he wrote that "at present,

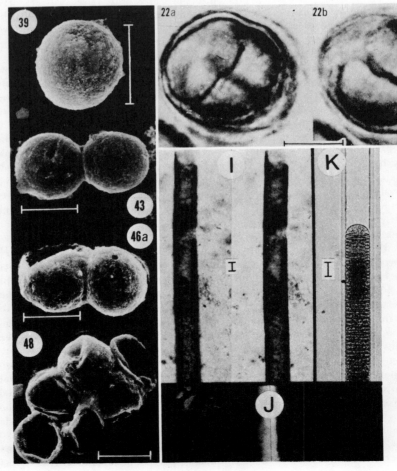

Well-preserved microfossils. Scanning electron micrographs (far left) and optical photomicrographs (top) show late (850 million years old) Precambrian fossils from Bitter Springs, Australia. Fossil algae, compared to modern blue-green algae (K), are late Precambrian, found by Yu. K. Sovietov in the Chichkan Formation in southern Kazakstan, U.S.S.R.

Eobacterium is probably best regarded as *suggestive* rather than *compelling* evidence of Archean life." (Archean is a subdivision of Precambrian time corresponding to the interval between 3.7 billion and 2.5 billion years ago.) As for the spheroidal microfossils that he and Barghoorn studied together more than a decade ago, Schopf now maintains that these "should not be regarded as constituting firm evidence of Archean life."

The range of sizes

The UCLA scientist is bothered by the range of sizes in which these spheroids seem to come. On the small end of the scale, some have been found with diameters of approximately a micrometer; at the opposite end, giant spheroids up to 193 micrometers in diameter have been discovered.

If all of these microscopic objects were biological, he feels, then billions of years ago there must have been some very odd biology at work. "Take a procaryotic cell, a modern bacterial cell," says Schopf, as he tries to explain his unease about the size distribution of the Fig Tree's objects. "Let's say it has a diameter of 100 microns [micrometers]. When that cell divides, it gives rise to two identical daughter cells, each with equal volumes and a diameter of, roughly, 80 microns. They then mature and become adult cells of about 100 microns diameter or somewhat more before they, too, split and produce a pair of daughter cells each.

"Okay, so let's suppose now that this colony of cells—parents, newborn daughters, adolescents and young adults—is killed. We measure the sizes of all these members and what we'd find would be a spread of maybe 60 or 65 microns diameter up to 115 or so. Because that's the way modern procaryotes divide and

Protagonists. Principal figures in the effort to unlock the secrets of Precambrian life include, from left, J. William Schopf, Preston Cloud and Elso S. Barghoorn.

grow—within a fairly well-defined band of sizes."

It is difficult for biologists to believe some other cellular reproductive process was operating, even billions of years ago. Could a 200-micrometer cell divide into two daughters, one of which was only a micrometer and the other 199 micrometers in diameter? Or could the original cell divide not into two daughters, but perhaps five equal-sized offspring? Or ten? Or twenty? "If that's true," says Schopf, "then it was a biology radically different from the biology of later times, a kind of cell division vastly different from the kind we know has been operable for a long, long time. It isn't very probable."

Alternative explanations are that the Swaziland spheroids (as the microfossils from the Fig Tree site and the surrounding South African region are often called collectively) are nonbiological in origin, as Cloud and some others suspect, or, some are willing to grant, a mix of biological and nonbiological objects.

And if the spheroids are of mixed biological and nonbiological origin, says Schopf, then the distinction between the two must somehow be explained. What processes imparted life to one class of chemical aggregates and not to another? Schopf notes that the suggestion has been made that the large spheroids are prebiological aggregates of organic compounds, formed in the chemical stew that is thought to have been the primordial oceans, and that the small objects are the first living cells. "I would bet that the little ones were biotic and the big ones abiotic," he hazards, "but I still think this has to be explained: how one came to be what *it* was and the other the way *it* was."

For his part, Cloud believes that the oldest microfossils for which there is *compelling* evidence are the deposits that he and his associates uncovered in Pokegama quartzite in northeastern Minnesota. (He concedes the probability that stromatolites more than three billion years old are *presumptive* evidence of simple algal forms at that time.) Seen under the electron microscope, the Pokegama microfossils have revealed long threadlike structures that Cloud identifies as the sheaths and bodies of filamentous Precambrian bacteria. The rock layers in which these were discovered occur beneath (in an older layer than) rocks contemporaneous with the nearby Gunflint strata, whose microbial biota Cloud calls "one of the 'golden spikes' in this business."

Schopf agrees that the Gunflint and Pokegama formations contain genuine fossils of Precambrian life. But he prefers, as the oldest clear-cut evidence of life, stromatolites found at Steep Rock Lake in Canada and at the Bulawayan Group in Rhodesia. These have been most recently dated as being between 2.6 billion and 2.7 billion years old (or between 2.5 billion and 3.0 billion years old; opinions vary). This choice is despite the fact that neither the Steep Rock nor the Bulawayan deposits contain structurally preserved fossils. They contain only the suggestive imprint of long-ago algal layers, any fossil algae that might have been there having long since disintegrated in the relatively porous limestone of these formations.

Few stromatolites contain fossils, Schopf observes. But the Steep Rock and Bulawayo layers, piled rather like ersatz potato chips stacked in a can, **are** suggestive of those stromatolites found elsewhere that do contain fossiliferous structures.

Oldest eucaryotes

But if Cloud and Schopf are reluctant to accept uncritically Barghoorn's claim to the oldest procaryotes, the Harvard professor is no less reluctant to accept their views as to the oldest eucaryotes. Schopf maintains that the microfossils he gathered from the Bitter Springs formation in central Australia in the early nineteen-seventies contain evidence of internal objects, objects that stamp these cells as eucaryotes; the formation has been dated as being about 850 million years old.

Aware that some of his colleagues regard these objects as the same sort of coalified dregs that have been seen in the Fig Tree and other microfossils, Schopf insists that there are subtle but distinctive differences in the size range, shape and distribution between the Bitter Springs specimens and others. To Schopf, these blebs represent evidence of membrane-bound nuclei or other organelles, characteristic badges of the more advanced, nucleated cell.

The size of the microfossils from Bitter Springs and some six other sites also figures prominently in Schopf's argument that the organisms they represent were eucaryotic: They are bigger than the procaryotes, even as the nucleated cells of today are considerably larger than those of bacteria and blue-green algae.

"All known assemblages of microfossils older than about 1.4 billion [years] are distinguished by cells with a mean diameter of about 5 microns," Schopf declares. "In contrast, microfossils younger than 1.4 billion tend to be bigger, with a mean of around 15 microns and some individuals going as big as 40 to 80 microns." Soviet scientists have reported a similar size jump at about the same time in microfossils unearthed within the U.S.S.R., he adds. No single piece of evidence is the clincher, Schopf notes, but the sum total of all this evidence suggests—at least to him—that nucleated cells first saw the light around 1.4 billion years ago.

Schopf's interpretation on this issue coincides with one first proposed by Cloud and his colleagues in 1969. Barghoorn, on the other hand, holds to the conventional date of 700 million years ago for the appearance of the first eucaryotes. Cloud, though in accord with Schopf on the age and the interpretation, rejects the Bitter Springs material. He contends, rather, that "the primary reference point for this key evolutionary event is the Beck Spring deposit in eastern California, a deposit indirectly dated at 1.2 billion to 1.7 billion years, and most probably about 1.3 billion years old."

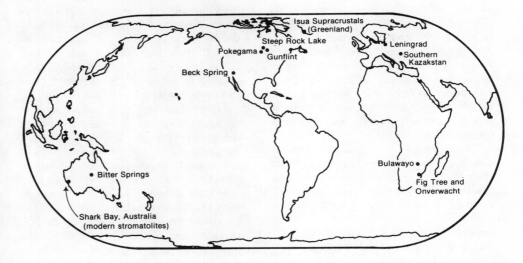

Principal sites. Microfossils contributing to the ferment in Precambrian biology have come from most of the world's continents.

The evidence found at Beck Spring consists of microscopic cells and branching filaments that range from 30 to 60 micrometers in diameter. Moreover, says Cloud, "Some of these filaments appear to have interior cross-walls, a feature that excludes the possibility that they could be the borings of smaller microorganisms and demonstrates a cell diameter that is most consistent with assignment to a eucaryotic microorganism." Cloud reasons, from evidence related to atmospheric evolution, that eucaryotes may even have originated as far back as 1.8 to 2.0 billion years ago.

Cells within cells

But how did this cellular silk purse arise from the sow's ear of procaryotic organisms? Boston University biologist Lynn Margulis has given new life to an old idea by hypothesizing that the first eucaryotes were colonies of symbiotic procaryotes. The colony enjoyed an advantage over individual procaryotes, and so survived.

Indeed, there is a great deal of indirect evidence to support this thesis. Even today there are one-celled organisms known which can assemble themselves into integrated colonies that function as organic wholes. Nor is it unheard of for a modern procaryote to invade a eucaryote and take up residence there as a symbiont.

Further, the sizes of procaryotes and organelles in nucleated cells are comparable, and the chloroplasts of photosynthetic eucaryotes bear a striking resemblance to some blue-green algae. Beyond that, some subcellular organelle types even have their own genes, perhaps a carry-over from the time, eons ago, when they were independent organisms.

It is quite possible, of course, that paleobiologists may never find traces of those first procaryotic communes, or the first procaryotes that became truly nucleated. "Fossils," says Schopf, "provide a record only of the products of the evolutionary process, not of the dynamic process itself." And where the process is evident, as in the case of simple organisms now inhabiting other cells symbiotically, there is no assurance that they truly mirror conditions billions of years old.

If, as Cloud, Schopf and others contend, eucaryotes first moved across the stage of life some 1.3 or 1.4 to 2.0 billion years ago ("The time control is rough," Cloud notes), why was there such a long hiatus before the multicellular metazoans appeared around 600 million years ago? Considering the enormous potential for variety and complexity which the eucaryotic cell possessed, it does seem like a long wait to cash in a winning ticket.

Not really, says Schopf. The potential of the eucaryote remained unrealized, he maintains, until eucaryotic sexuality arose (the first eucaryotes would have been asexual), and then only after some pressure was applied to it. That pressure may have come eventually in the form of predation.

It is an ecological argument. Ecologists know, for example, that a population of organisms not exposed to any pressures will be stable and limited to a few dominant types. These dominant types tend to monopolize their environment and its resources almost to the exclusion of other, different species.

Introduce a predator into this seemingly idyllic situation and a curious, even paradoxical, outcome develops: Every species benefits; the predator thrives because of the great availability of prey, and the prey is reduced but not eradicated; the dominance of a few prey species is cracked, providing niches that other prey species can occupy.

These formerly constrained species now begin to expand and, in turn, attract their own adapted predators. The net result of all these interactions, which ecologists call cropping, is "a sort of ecological feedback system... that results ultimately in an increase of biological diversity at all levels of the food web," according to Schopf.

In his view, based largely on studies by Steven M. Stanley of The Johns Hopkins University, the Precambrian environment probably lacked predators until somewhere around 800 or 900 million years ago. "During the late Precambrian," Schopf wrote in a 1975 paper, "as sexual eucaryotes became increasingly abundant and diverse, it can be surmised that many new types of microorganisms first appeared."

Among these new creatures, Schopf speculates, were some eucaryotes which, lacking the ability to photosynthesize the nutrients necessary for their survival, obtained what they needed by being "omnivores, feeding on organic detritus and small cells."

The effect of predation would have been dramatic. It could, Schopf and others who subscribe to this view believe, have led to a marked increase in the biological diversity and ecological complexity at all levels of the food chain. Moreover, it could culminate, in Schopf's words, "in the appearance and rapid diversification of macroscopic croppers [invertebrate animals] and co-evolving macroscopic 'croppees' [advanced, multicellular algae]." In other words, the second act—the Cambrian—in the drama of life was about to unfold.

The atmospheric connection

Nothing ever being quite as simple as it might seem, there are still many unresolved questions surrounding the evolution of life. One such question concerns the earth's atmosphere and its gaseous composition.

"Due in large part to volcanic outgassing," says Heinrich Holland, a Harvard professor of geochemistry, "the early atmosphere was heavily [made up of] carbon dioxide, with perhaps no more than one percent free oxygen. That was between two and three billion years ago." But analysis of rocks and minerals indicated that this changed around two billion years ago, as photosynthetic organisms devoured the carbon dioxide and released oxygen.

Some scientists have proposed that metazoans could not have developed until there were sufficient levels of oxygen in the atmosphere, both for these advanced organisms to breathe and to shield them from the lethal ultraviolet rays of the sun. This critical level was, according to one view, reached at about the time the Cambrian era began—some 600 million years or so ago—and, in fact, was the event that brought about the Cambrian explosion of life forms.

Schopf agrees that an oxygenic atmosphere is necessary for metazoans to exist, but he contends that it was also necessary for their eucaryotic predecessors. He argues that considerable amounts of free atmospheric oxygen had already built up around two billion years ago, citing a variety of geologic data to support his claim. As in other areas relevant to paleobiology, the last word on this issue is yet to be heard.

Paleobiology is a field in ferment; it is not at all unusual for there to be wide divergence of opinions and interpretations among the people active at such a stage in any discipline. Each of these researchers has particular investigations which he intends to carry out and which will yield, each hopes, data that will resolve hanging questions.

Schopf, winner last year of the NSF's Alan T. Waterman award, plans to use the proceeds ($50,000 a year for each of three years) to enable him to assemble an interdisciplinary team, including geologists, chemists, biologists, paleobiologists and others, from universities around the world. Only such a team, he holds, can successfully attack the array of unanswered paleobiological questions directly. The prize money, he says, will be used for travel and salary expenses of the team members.

"There are still a lot of gaps in the [paleobiological] record," Schopf says. "With the right people, I think we'll be able to fill them in. But in the end, it comes down to the material that you have to work with. The rocks," he says, "are the court of last resort." •

The research reported in this article is supported largely by the Systematic Biology and Geology programs of the National Science Foundation.

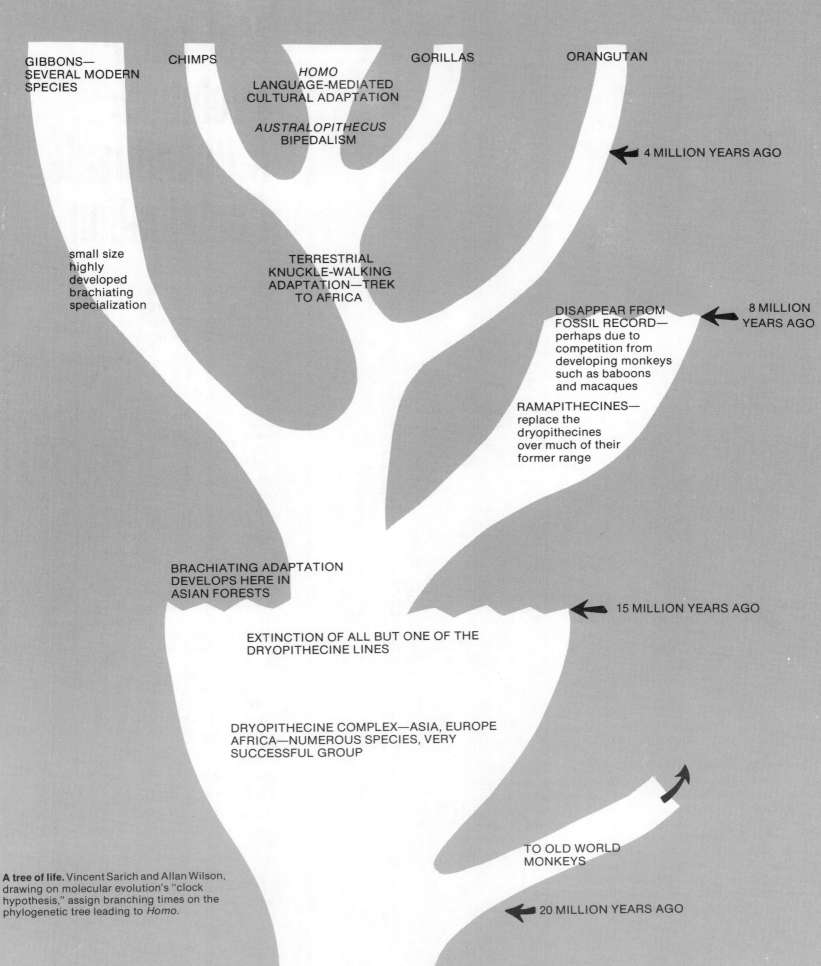

A tree of life. Vincent Sarich and Allan Wilson, drawing on molecular evolution's "clock hypothesis," assign branching times on the phylogenetic tree leading to *Homo*.

Molecular Evolution: A Quantifiable Contribution

Changes in genetic material and proteins help time the stages of species evolution.

The epic of the emergence of the genus *Homo* commonly takes the form of a metaphor: the "tree of life." To biologists, anthropologists paleontologists and others, the tree is a branch-by-branch genealogical reconstruction that seeks to link to its common trunk every form of life, all the way back to the primordial common root.

Needless to say, the story is not complete. Large gaps remain in our knowledge of the phylogenetic tree, gaps relating not only to the order of branchings but even more to the times in all those millennia past when the divergences occurred. For investigators of the evolutionary process, a principal goal continues to be the fulfillment of the metaphor: the completion of the tree.

Why the gaps? Why the incomplete tree? Simple: lack of sufficient data from which to reconstruct the full phylogenetic profile.

What evidence there is has come traditionally from two major sources: comparative morphology and the dating of the fossil record. The former seeks to establish genealogical relationships among extant as well as ancient species on the basis of anatomical similarities; the latter has long been the only way to get an idea of temporal relationships—the times of divergence, the branching dates at which one lineage became two.

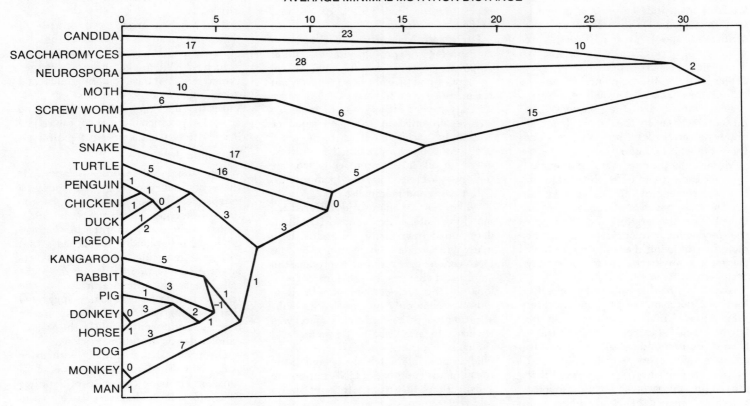

Molecular phylogeny. Differences in amino acid sequence of cytochrome c produced this computer-based evolutionary tree. Numbers on the branches relate to nucleotide substitutions in DNA necessary to produce the amino acid differences between species.

Walter Fitch, Emanuel Margoliash

Phylogenetic profiles emerge from such data, but they are often rife with unspecified inference. Different investigators, looking at the self-same evidence, have often produced widely differing constructs. Branching times have been a particularly sticky point.

"In reality," says Vincent Sarich, a biochemist-cum-anthropologist at the University of California at Berkeley, "there's only one real evolutionary history, only one true tree. Ultimately all available data—from the fossils, from modern anatomy—must be consistent with this one true tree. If you get discordant answers then something is wrong."

A quantifiable approach

Over the last two decades, in the wake of striking advances in molecular biology, a new scientific discipline has been emerging with the potential to damp the discord. Called molecular anthropology or, more broadly, molecular evolution, it devolves from the same Darwinian postulate that first set scientists to examining anatomical similarities among species for clues to the proper ordering of phylogeny.

The Darwinian precept: Evolution is forever divergent. When species split off from a common ancestor, they grow more and more different over time, and the more two species differ, the more distant must be their common lineage; conversely, the more alike two species are, the more recent must be their common ancestry. The principle applies not only to morphology but to the stuff of heredity itself—the large deoxyribonucleic acid (DNA) molecule and to the proteins manufactured on instructions from codes enfolded in DNA's double helix. Thus, amino acid components of like proteins in different species are different; the degree of difference can be a measure of the evolutionary distance between the species. For example, the basic cellular protein, cytochrome c, is the same in humans and chimpanzees but varies by 44 out of a total of 104 amino acids from that of *Neurospora*, a fungal organism widely used in genetic research.

Investigation of evolutionary history at the molecular level offers two significant advantages over morphological comparisons. One is that the information is highly quantifiable. With molecules, scientists can tote up such differences as the number of amino acid substitutions, leaving little room for subjectivity. Additionally, molecules allow organisms that are very distant phylogenetically to be compared. Comparative anatomy can't do much with organisms as diverse as *Neurospora*, fish, birds, insects and primates, but there are proteins they all have in common.

In the beginning

The idea of using molecules to study phylogeny dates back to a paper published in 1902 in the *British Medical Journal*. British serologist G. H. F. Nuttall suggested then that a comparison of blood proteins could reveal a good deal about genealogical relationships among primates. The proposal followed briefly the 1900 "rediscovery" of Mendelian genetics.

It took the cracking of the genetic code more than a half-century later, however, to fill in the outlines Nuttall had sketched—to explain what the evolutionary message in proteins was and why proteins were such a profound link to ancient human ancestry.

At the root of it all are the genes, relatively short segments of the long, double-stranded DNA molecules. Each strand of the double DNA helix represents a linear code made up of four kinds of nucleotide bases. The information contained in the code directs the synthesis of specific proteins. Proteins are made up of long chains of amino acids. The specific properties of each protein are determined by the sequence of amino acids in the chain. That sequence, in turn, is determined by the sequence of nucleotide bases in the DNA. Thus an organism can be described as the sum of the proteins it manufactures.

Mutation is the raw material of evolution. At the protein level, this can mean the replacement of one amino acid for another, and such differences can be tallied. The greater the tally, the larger the phylogenetic distance. Hence, the molecules of living species preserve their place on a phylogenetic tree in readily decipherable form.

Armed with this molecular insight, Morris Goodman, an immunologist at Wayne State University in Detroit in the late nineteen-fifties, resurrected Nuttall's idea, using immunological techniques to compare like proteins from different species. He published results in 1962 that proved the approach valid while it toppled a well-entrenched belief about primate evolution: the belief that chimpanzees and gorillas were much closer kin to each other than either was to humans, that those anthropoids, as the morphological evidence appears to testify, continued to share a common ancestral stem long after the hominid branch, leading uniquely to humankind, had split off from the hominoid trunk to which all three belong.

The molecules told a different story. They revealed that humans are as close genetically to the gorilla and chimpanzee as these two apes are to each other. Moreover, gorillas, chimpanzees and humans were found to be closer to one another genetically than any of them are to orangutans or any other primates. All three must have developed from their common ancestor at about the same time. In the ensuing years, a variety of methods for assessing molecular differences have come into play; all confirm Goodman's original observation.

A molecular tree

Goodman's research helped put the discipline of molecular evolution on the map. A landmark paper published in 1967 glued it there. The report, by Walter M. Fitch of the University of Wisconsin and Emanuel Margoliash of Abbott Laboratories in Chicago, demonstrated the feasibility of using molecular differences to construct phylogenetic trees. It also offered a rigorous mathematical procedure for linking each of the taxonomic branches.

Fitch and Margoliash worked with cytochrome c. A protein vital to cellular respiratory processes, its amino acid sequences had already been worked out for many species. In each case, however, no matter what the species, the molecule always contained an unbroken sequence 104 amino acids long; each of these positions could be examined for differences. In all, 20 different cytochrome c molecules were compared—from insects, fish, turtles and mammals.

Whenever two amino acids differed at a matching site, Fitch and Margoliash determined how many mutations, how many nucleotide substitutions in the DNA codon, were required to produce the amino acid changes. In some instances, a look at the well-defined genetic code indicates that only one nucleotide substitution in a three-nucleotide codon was required to bring about the amino acid change; in others, at least two. As a guiding principle, Fitch always opted for the minimum number of DNA nucleotide substitutions to account for any amino acid replacements.

In essence, this approach was a direct application of the principle known as Occam's Razor, which proposes that the number of assumptions required to explain an event be kept to as few as necessity demands. This criterion of economy, and procedures for its use in molecular evolution, has since come to be called the "maximum parsimony principle" and has been widely adopted by many phylogenetic tree builders.

Through their exploration of differences, Fitch and Margoliash managed to derive a series of numbers. The larger the number separating any two species, the more remote their common ancestry and the farther apart their branch points.

The tree they drew, using only the data from cytochrome c, proved to be nearly identical to the tree produced on the basis of morphological data. In effect, Fitch and Margoliash showed that by studying the product of merely a single protein-coding gene they could come up with a solution much like that obtained by studying morphology, the product of thousands of genes.

Since then, a number of other molecules have been subjected to extensive cross-species comparisons. No matter the molecule or the technique for measuring molecular differences, the branchings of the derived phylogenetic trees have been in general agreement. (See "Methods for matching molecules," accompanying this article.)

But branching order is only one part of the evolution story. What about the *when* of evolution? When did the tree sprout branches? When did lineages diverge? When, for instance, did hominids and other hominoids begin to go their separate developmental ways? The answers are crucial, but the fossil record has yet to provide a clear-cut solution.

Detecting the tick

With the fossil record ambiguous, is there another way to go? Is there an independent line of inquiry to solve the time-of-divergence dilemma? In 1962, Linus Pauling and another biochemist, Emile Zuckerkandl, proposed that biological molecules, established as arbiters of the sequential issue, might be called in as the ultimate arbiter of the temporal issue as well. They suggested that proteins evolved at a fairly steady rate. Such uniformity could be the basis of a molecular or evolutionary clock.

The proposal met initially with utter skepticism. But over the past decade a considerable body of empirical data has been amassed to lend credence to the revolutionary idea that protein ticks—amino acid replacements—came at reasonably constant intervals. Ultimately, the notion grew that these ticks could be used in pinpointing dates of phylogenetic divergence.

Not that the ticks are metronomic. They aren't. The clock is not of the same order as say, a wristwatch, in which each and every tick comes at a precisely regular interval. Rather, the proposed evolutionary timepiece is probabilistic, or stochastic, in a way that's akin to radioactive clocks. The principle of such a clock is that a certain number of ticks (radioactive decay of atoms; mutations) are likely to occur within a given time period, but one can predict precisely neither when the next tick will tock nor which atom or protein will alter next. Nevertheless, as is evident from success with the analogous techniques of radioisotope dating, the average is sure enough for calculation of elapsed time to be made. By counting the "ticks" (the differences in the amino acid sequence in equivalent proteins from any two creatures) and establishing the frequency of the ticks, it should be possible to determine just how long ago species had diverged from a common ancestral stem.

The molecular clock

Much of the evidence supporting the molecular clock hypothesis has been gathered by Sarich and his Berkeley colleague, biochemist Allan Wilson, looking principally at the blood proteins albumin and transferrin. Because of the enormous size of these molecules (an albumin molecule is made up of a string of 580 amino acids), direct amino acid sequencing was out of the question. Instead, Sarich and Wilson perfected an immunological method for measuring indirectly, but quantitatively, amino acid differences.

The investigators discovered that, since the primates first emerged, each of the major primate lineages (human, ape, orangutan, gibbon, Old World monkey and New World monkey) had accumulated very similar numbers of amino acid replacements in albumin and transferrin. The determination was made by comparing these proteins from blood from each of the primate lines with equivalent proteins from an "external reference" or non-primate species. The California scientists ran comparisons on a multitude of species all across the vertebrate spectrum. Invariably, the results were the same: a like number of changes in the lineages over an identical span of evolutionary time. The number of ticks along each of the lines, since they originated, was very similar. Time alone appeared to be the primary propellant of protein evolution. Wilson and Sarich's conclusion: "The data definitely document the existence of the clock."

Their clock, based on albumin-transferrin substitutions, does not apply to other molecules, which evolve at different rates. Cytochrome *c*, for instance, ticks off a one-percent change every 20 million years. Hemoglobin, however, which seems to be more tolerant of change, substitutes amino acids at the rate of about one percent in just six million years. But whatever the molecule and whatever the ticking rate, says Sarich, the derived divergence dates have been found to be in good accord.

Setting the clock

What remained was to "set" the clock—to translate the ticks into actual times of divergence in millions of years before the present. Clock calibration required an external time reference. One such source: fossils. Sarich and Wilson settled then on a base-point 30 million years ago as a conservative estimate for the divergence of the Old World monkey and ape lines. (Subsequent work suggests that that monkey-ape split was more like 20 million years ago, Sarich notes.)

With the clock—the linear relationship between genetic distance and time of separation—established by "relative rate" tests, says Sarich, and a base time selected, it was then "simply a matter of choosing that straight line which would make the largest number of paleontologists least unhappy."

Then Sarich and Wilson produced a bombshell: They concluded from their data that the divergence of African apes and man came barely five million years ago, give or take two million years.

By contrast, paleontological estimates had not too long ago placed that divergence as far back as 30 million years. And paleontologists are still not willing to come far to this side of some 14 million to 20 million years ago as that key branching time. Anything much less, they contend, would leave far too few millennia for all the changes between early hominids and *Homo*. A five-million-year branching date, for instance, would all but knock out of the running a major contender for the title of first hominid: *Ramapithecus*. The fossil record first from Northeastern India and Pakistan and then from such places as Kenya, Turkey and Macedonia, say *Ramapithecus* dwelt on the planet at a time too far back—14 to 8 million years ago—to be in the clock-based hominid line. And even though paleontologists like Elwyn Simons of Duke University and David Pilbeam of Yale differ over *Ramapithecus*'s credentials, even Pilbeam still looks to some of that hominoid's Miocene contemporaries for the roots of humanity. But for Sarich and Wilson, they are all—at between 14 and 8 million years back—far too early.

The clock's divergence hypothesis, however, does put the hominid/hominoid branching point neatly some million years before what is now regarded by paleontologists as the earliest known hominid: the 3.8-million-year-old *Australopithecus afarensis*, a biped associated with fossil footprints found in the Laetolil beds of Tanzania and with somewhat younger fossil remains from the Afar triangle of Ethiopia.

A. afarensis is now widely accepted as the earliest known hominid. But nobody is suggesting that it is the first hominid.

Unfortunately, however, the hominid fossil record between *A. afarensis* and *Ramapithecus* is so far a tantalizing blank; other than the molecular data, no easy resolution is in sight for the paleontological controversy over hominid antecedents over that four-million-year void.

Inevitably, the clock hypothesis—threatening as it seems to some orthodox tenets—has unleashed sharp debate and criticism. The clock advocates, data in hand, have offered counters to most of the points raised against them. For example:

Point...

The clock theory says that mutations—substitutions in the DNA codon leading to amino acid replacements in protein—occur at a fairly constant rate in all species. Wouldn't that argue for reasonably similar rates of evolution? Yet evidence to the contrary seems to abound. Compare, for instance, the morphological diversity among such mammals as bats, cats, whales and people with that among frogs. Despite the fact that frogs have dwelt on the planet much longer than placental mammals, they are—all 3,000 species—very much like each other in appearance. So alike are the frog species that zoologists have placed them in just one phylogenetic order, Anura, whereas placental mammals occupy at least 16.

Or take the example of apes and humans. The same evolutionary clock presumably ticks in both. And yet, even though they shared a common, apelike ancestor, the human lineage has undergone significantly more phenotypic (morphological) change. Obviously, evolution on the level of the whole organism has proceeded in anything but a regular, clocklike fashion.

...counterpoint

The evidence merely shows, Sarich and Wilson counter, that changes in proteins need have no bearing on gross morphological change. Wilson points out that two frog species similar enough to be included in a single genus can differ from each other at the molecular sequence level by as much as a bat does from a whale. Similarly, an examination of the higher primate lineages indicates that, in spite of an accelerated rate of phenotypic evolution, humans do not show any signs of stepped-up protein substitution. "The ticks of the molecular clock say nothing about how different species will be in outward appearance," says Wilson. "Sequence evolution is a function of time and proceeds as rapidly in morphologically conservative creatures as in those which have changed radically."

If so, then how are the widely diverse rates of morphological evolution to be explained? According to Wilson, that's under the control of genes involved not in the blueprinting of proteins but rather in regulatory functions (i.e., those governing the rates and timing of protein synthesis). Basic studies in bacteria and algae seem to lend preliminary support to this postulate. Wilson predicts that, when scientists become adept at identifying and sequencing those portions of the genome concerned with regulation, they will uncover mutational rates that correlate strongly with rates of organismal change.

Point...

Other investigators are far less skeptical than are Sarich and Wilson of the paleontological estimates that push the ape-human split much further back in time. One such is Morris Goodman, the Wayne State immunologist whose molecular data demonstrated the close evolutionary cousinship of hominids to gorillas and chimpanzees. Goodman says outright: "I do not reject paleontological evidence indicative of ancient splitting times in Hominoidea." For example, it does not seem inconceivable to him that the split between gibbons and other hominoids occurred in the range of 20 million years ago or so and that the 8- to 14-million-year-old *Ramapithecus*, or a contemporary, could be in the line of human ancestry.

Goodman poses one of the strongest challenges to the clock hypothesis. His data argue that clock-derived divergence times could result from misinterpretation of molecular events as they really happened in the course of the evolutionary millennia.

For one thing, he contends, there was a significant slowdown in protein evolution in the higher primates. Further, as he reads them, the molecules tell Goodman that there is very little regularity in protein change anywhere on the phylogenetic tree. Instead, evolution has proceeded by dint of nonuniform starts and stops, periods of rapid protein evolution followed by periods of evolutionary quiescence. What looks like a clock, he says, is only the "illusion of constancy," a consequence of averaging essentially irregular events so as to make it seem as if a clock was ticking. To Goodman's way of thinking, molecular evolution is good primarily for resolving one major phylogenetic issue: getting the branching order right. "But don't count on the divergencies among protein molecules to provide accurate dates," he declares.

Goodman's position stems from evidence derived from multispecies analysis of hemoglobin, the oxygen-carrying blood protein. ("Why isn't he addressing transferrin?" Sarich challenges.) In one series of studies Goodman discovered that the epoch 500 million to 400 million years ago, when birds and mammals were in the preliminary stages of emergence, was a special period of vertebrate ferment. Hemoglobin underwent enormous change, as molecules from representatives of species surviving from that epoch testify. Moreover, the specific sites on the molecule where the changes occurred during this dramatic transitional

---

Methods for matching molecules

Evolutionary relationships have long been established by anatomical comparisons among living species and by examination of the fossil record. Now macromolecules—DNA and proteins—are helping to refine the construction of the tree of life. The principle is simple enough: Evolution, a process marked by mutation, is divergent. The further back in time lineages split from a common ancestor, the greater the degree of genealogical distance. This evolutionary apartness is reflected in the molecules. Amino acids from similar proteins of different species are different, and the degree of difference corresponds to the degree of divergence between them.

Four methods are currently available to make these evolutionary distinctions. The differences devolve on individual changes—in the amino acids or in the DNA nucleotides that code for the protein. As such, all the methods—electrophoresis, immunology, amino acid sequencing and DNA hybridization—measure countable, unit differences which can then be apportioned additively onto the branches of a phylogenetic tree.

Electrophoresis. Proteins are placed on a gel, which is then subjected to an electric current, usually for a few hours. Each protein in the sample migrates through the gel in a direction and at a rate that depends on its molecular size and electric charge. The technique is so sensitive that it can identify proteins that differ by a single amino acid, as long as the single substitution produces a change in the protein's net electrical charge.

Amino acid sequencing. Proteins derive their properties from the order in which their building blocks, the amino acids, are strung together. The process of blueprinting the exact sequence is laborious and expensive. It takes about an hour to identify each amino acid in the protein chain. Hence, few attempts are made to sequence proteins that are made up of several hundred amino acids. Much of the sequencing to date, in fact, has been done by biochemists interested not in problems of evolution but in the mechanism of protein function.

Immunology. For large proteins—albumin, for instance, which has 580 amino acids—sequencing is out of the question. What has evolved instead is an indirect method for establishing amino acid differences. In essence, it calls for injecting rabbits with the test protein from one species, say humans. The rabbits produce antibodies. These antibodies are then reacted with the same protein from other species. The intensity of the reaction is a measurable, reliable index of the number of amino acid positions at which the proteins are similar.

DNA hybridization. Also called DNA annealing, the procedure is to the genes what the immunological technique is to proteins, an indirect method for determining the extent to which species are identical at the level of the DNA nucleotides. In this case, a hybrid, double-stranded molecule is formed by allowing single-stranded DNA from two different species to react (anneal) with one another. Then a measurement is made of the temperature—the "melting point"—at which the hybrid DNA separates again into two strands. The more dissimilar the nucleotide sequences for equivalent genes—i.e., the further away the species are genealogically—the lower the melting point.

Which method to use depends largely on the phylogenetic problem under investigation. Electrophoresis, for instance, is very useful for discriminating among similar proteins from closely related organisms.

On the horizon are techniques for sequencing DNA nucleotides directly. This point-by-point analysis of the genes themselves is expected to produce more phylogenetic information per unit than all the other techniques. The development of DNA sequencing results from the recent discovery of restriction enzymes, each of which recognizes and cuts one and only one specific sequence of six nucleotides on the genome. Such precision will go a long way toward giving anthropologists, molecular and otherwise, data they sorely need.●

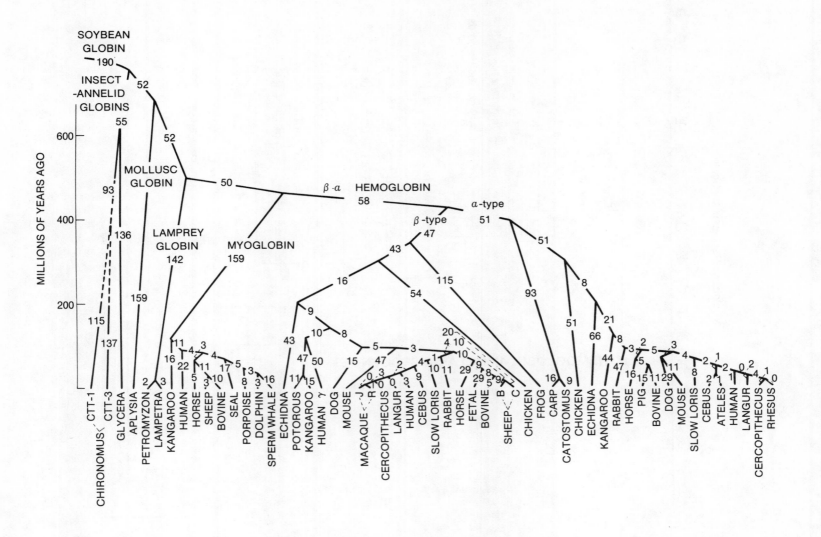

Variable rates of change. The number of nucleotide replacements (numbers of branches) that separates species, according to differences in globin protein, and the time of their branching from a common stock as derived from fossil evidence. By this index, evolution accelerated between 400 million and 500 million years ago in the descent of vertebrate hemoglobin genes.

Morris Goodman

age coincided with the sites responsible for the subsequent advance in function. One outstanding example, says Goodman, was the "especially rapid rate of protein change in the vertebrate stem leading to the early tetrapods, when the evolutionary path to a more mobile, active mode of life required that these advancing vertebrates have a hemoglobin which could rapidly unload oxygen."

According to Goodman's calculations, the average rate of nucleotide replacement in the DNA hemoglobin codons (using Fitch's maximum parsimony principle) was 18 or 19 every 100 million years, for the past 500 million years. Yet between 500 million and 400 million years ago, during that special period of evolutionary stir, the rate change was five to ten times as high, amounting to an average of more than 100 nucleotide replacements. Other proteins, such as cytochrome and carbonic anhydrase (a protein involved in carbon dioxide transport) have been put through similar kinds of analyses. They too show an irregular pattern, says Goodman, marked by bunched-up bursts of change followed by a marked slowdown, which has become especially evident in the higher primate lineages.

Goodman attributes the acceleration-deceleration evolutionary pattern to a process called anagenesis, in which he incorporates "that form of progressive evolution which increases the level of molecular complexity within organisms in such a way that the organisms have greater independence from and control over their environment." In other words, as lineages advance—Goodman calls it "positive evolution"—they are much less vulnerable to environmental change. At the same time, as the molecular architecture of a protein grows more complex or function-specific, it also becomes much less tolerant of change; further mutations will prove deleterious and fail to survive.

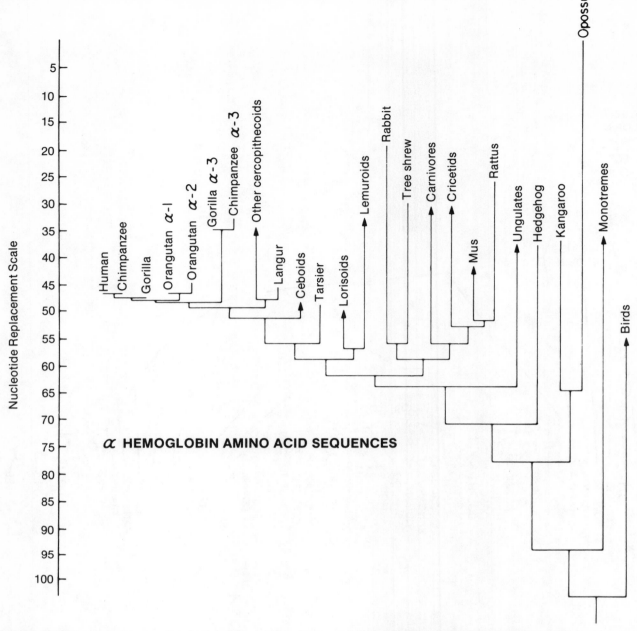

Distance between species. The bird-mammal portion of a "parsimonious" genealogical tree constructed for 70 alpha hemoglobin sequences. Scale suggests the sharp decrease in fixed mutations after anthropoids branched off from common ancestral species. For Hemoglobin-1 and -2, humans, chimpanzees and gorillas are more like each other than any of the three are like orangutans.

Morris Goodman

In his scheme, accelerated protein evolution comes when new environmental conditions exert strong, selective pressures for new adaptations at the organismic and molecular levels. Anagenesis then acts to preserve these improvements, thereby strengthening the resistance of higher organisms to variations in the environment. The net result: a trend toward a slowing of rates as lineages advance.

...Counterpoint

And what say clock proponents to all this? Do Goodman's data blow the works? Not at all, says Wilson. First off, he thinks Goodman may put too much stock in some of the fossil data he used to pinpoint his fast-slow evolutionary epochs, reflecting what Wilson calls "a touching faith in what the paleontologists have to say." But there's a stronger counterargument, too.

Sarich and Wilson have devised and run a test which, they contend, refutes Goodman's thesis that the clock is illusory. Their "relative rate" test compares the number of amino acid sequence replacements that have accumulated along any two lineages since their divergence from a common ancestor. The test requires no need for knowledge of the specific time of divergence; only relative amounts of change are under scrutiny. The test asks: Is the number of mutations along one lineage statistically identical to that along the other? If slowdown of evolutionary rate has occurred, in accordance with the process of anagenesis, then this should be evident when a more evolutionarily advanced form is matched against a less "progressive" species.

Sarich and Wilson compared eight different proteins, including hemoglobin and carbonic anhydrase, from higher primate lineages with those from less advanced, Old World monkeys. Their conclusion: "The results give no support to the idea of a molecular evolution

RELATIVE DISTANCES AMONG THE PRIMATES

Groups compared	Albumin plus transferrin immunology	DNA	Time of separation (millions of years ago)
Man—chimp—gorilla	0.12	0.14	4—5
(M,C,G)—orangutan	0.25	0.3	9—11
(M,C,G,O)—gibbons	0.29	0.36	11—13
(Apes, man)—Old World monkeys	0.57	0.60	20—22
Old World—New World higher primates	1.0	1.0	35—38
Higher primates—prosimians	2.0	---	70—75

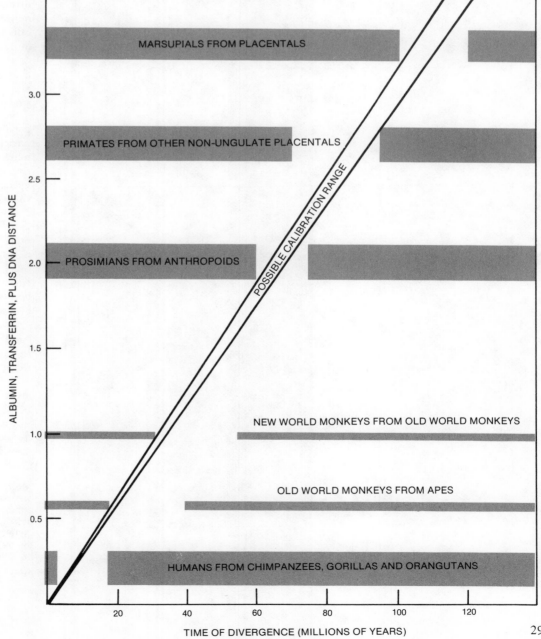

Molecular clock. The linear relationship of the molecular disparities between two species to their divergence time is used to demonstrate the clock hypothesis. Relationships are drawn from albumin, transferrin and (for the lower three divergences) DNA. The scale is a proportion of the difference (1.0) measured between the higher primates of the Old and New Worlds. The left-hand bars show the acknowledged presence of derived species; right-hand bars show the presence of common source lineages. Available straight lines in the gap between represent the "possible calibration range" by which molecular differences can establish divergence times.

Victor Sarich, Allan Wilson

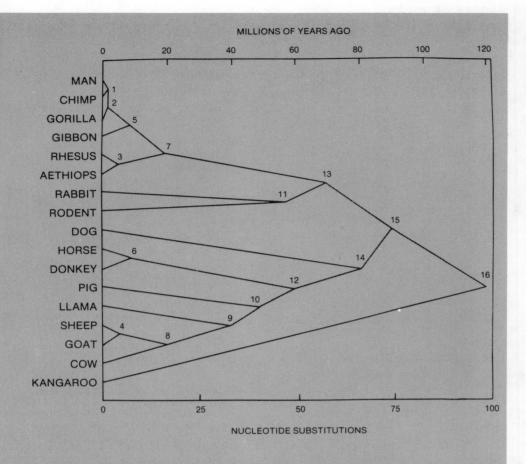

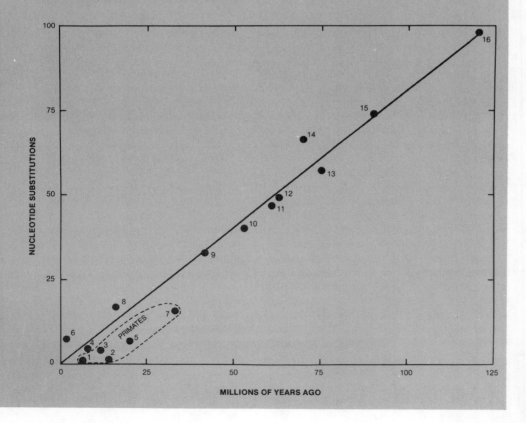

slowdown specific to the lineages leading to humans and African apes." Wilson sees no reason to doubt the clock hypothesis, nor to retreat from its five-million-year-old divergence date.

Neutralist, selectionist, empiricist

Compelling arguments from both fronts—but who's right? Is protein evolution clocklike or not? The answer, of course, is of more than passing interest.

In philosophical terms, the issue has been described as a debate between "neutralists" and "selectionists." The neutralist position holds that most substitutions that take root in the population are neutral from the standpoint of evolutionary advantage. The altered protein continues to function because the changes involve only certain amino acids; the substitute has similar chemical properties, so the conformation and function of the proteins are altered very little. Change in protein, say the neutralists, is primarily due to what is called the random drift of chance mutations; ultimately, environmental or competitive stresses will push one changed form out of the race, while another—and its protein configuration—survives.

Neutrality seems to be an inherent condition for the clock hypothesis, and Sarich and Wilson have been tagged as neutralists. It is a label they shun, styling themselves instead as empiricists, investigators concerned more with data than ideology.

The selectionists, a side to which Goodman offers ready allegiance, contend that the Darwinian principle of natural selection applies to proteins as much as it does to the whole organism. Any fixed change in protein happens because it is adaptive—it offers some evolutionary advantage.

Once again: uniform or nonuniform? Neutral or selectionist evolution? Was the issue resolvable? Into the fray stepped Walter Fitch, this time in collaboration with Charles H. Langley of the National Institute of Environmental Health. In 1975, they undertook what has proven to be the most rigorous examination of the clock hypothesis to date. The question they addressed was: Does protein evolution occur at a uniform rate? Put another

"Forcing" the clock. Nucleotide substitutions compared among 17 species and plotted against times of divergence according to the paleontological record (above) show generally straight-line agreement between the two (below). Verdict: a molecular clock, but "sloppy."

Walter Fitch, Charles Langley, Leigh Van Valen

way: To what degree are protein changes governed by time alone?

Toward harmony

For their line of inquiry, they chose the "null hypothesis" test that assumes, for the purpose of the test, that the theory under investigation is valid. The predictions derived on the basis of the theory are then checked against what actually happened. If the theory works, then its predictions should coincide with actuality.

The investigators studied seven different proteins from 17 widely divergent mammals, including humans, gibbons, rodents, dogs, horses, cows and sheep. As different classes of protein evolve at different rates, each class was assessed independently. Changes in protein along each mammalian lineage were determined by comparison with protein from a nonmammalian reference species, a marsupial.

Fitch and Langley then "forced" the clock on the data, seeking a distribution that would look clocklike and locating phylogenetic branching points at uniform intervals.

As an arbitrary reference point, they chose a time "99 nucleotide substitutions from today ago," a time when marsupials and placental mammals diverged. Then they calculated, based on the radioactive clock model, the expected substitutions between this reference point and the various branch points of the phylogenetic tree.

This was the "what-the-theory-says-should-have-happened" part of the test, if molecular evolution were as clocklike as the decay of radioisotopes. The "what-actually-happened" data came from an actual measurement of the nucleotide substitutions.

The verdict: If there were a molecular clock, it wasn't as clocklike as a radioactive clock. It was, in a word, "sloppier"; the average interval between ticks was a matter of significant fluctuation. The variability—or variance, as statisticians call it—was found to be about twice that seen in the radioactive clock. According to Fitch, this variability was apparently a direct consequence of constraints imposed by natural selection. It upset the neutralist postulate that virtually all protein replacements are neutral events. The sloppiness suggests that, while many protein changes may be neutral, a sizable fraction are not.

Did "sloppy," then, spell doom for the clock hypothesis as a means of pinpointing the dates of evolutionary events with any degree of confidence? Not at all, says Fitch, as a corollary test demonstrated. Fitch asked Leigh Van Valen of the University of Chicago to make an independent paleontological determination of times of divergence for the mammalian test species. There was little concern about the validity of these estimates, since much of the mammalian fossil record, other than that relating to higher primates, is excellent and not a matter of debate.

Against these paleontologically derived times, Fitch plotted the nucleotide substitutions. The result was a straight line, an indication that the fit or match-up was surprisingly good. It means that, says Fitch, given enough molecular data, the clock could be used to get good estimates of divergence.

To empiricists Wilson and Sarich, proof of the existence of a clock, sloppy or otherwise, validates an "astonishingly powerful tool" for broaching long-elusive riddles related to the evolutionary saga. And they are already at work, seeking the extra data necessary to clean up the sloppy clock. This can be accomplished, Wilson explains, by measuring more ticks over a given interval.

One way to get more ticks is to move from sequencing proteins, made up of hundreds of amino acids, to sequencing segments of DNA having millions of nucleotides strung out along their lengths. As it happens, new techniques for sequencing DNA directly have been coming along fast. Not surprisingly, Wilson has shifted the entire focus of his biochemical laboratory toward DNA investigation.

In sum: Issues surrounding the molecular clock are in dispute; a molecular basis for the study of evolution is not. Molecular evolutionists on both sides of the clock debate—or neutral—hold equally to one verity: macromolecules are opening a new chapter in the long-standing effort to complete the narrative of evolution. They are providing a quantitative way to help realize an important dream of the scientist who may be said to have planted the phylogenetic tree: Charles Darwin. In a letter to his friend Thomas Huxley, another ardent evolutionist, Darwin wrote: "The time will come, I believe, though I shall not live to see it, when we shall have fairly true genealogical trees of each great Kingdom of Nature."●

The National Science Foundation contributes to the support of research discussed in this article principally through its Systematic Biology Program.

The World's Great Dyings

by Arthur Fisher

Some subtle geochemistry has identified a bona fide interplanetary catastrophe with the mass biological extinction on earth 65 million years ago. The nature of the biological event is still a subject of some controversy.

"Certainly no fact in the world is so startling as the wide and repeated exterminations of its inhabitants," wrote Charles Darwin after pondering the disappearance of many large mammals from the pampas of South America. Today, long after he wrote those words, scientists are still advancing theories to account for such "wide and repeated exterminations." (Today's paleontologists call them mass extinctions.) The proposed causes, some of which stand up better than others, range from titanic lunar eruptions to human predation to nothing special. These propositions, and the vanishings they seek to explain, are at the hub of a debate over the true course of life on earth.

The debate is an intellectual brawl that began late in the eighteenth century. It is still going on. One camp holds that the geological history of the planet, together with the record of past life as revealed by fossils, can be explained by events occurring very gradually over immense periods of time. The other insists that the pattern of awesome changes in the earth's surface, coupled with apparent disappearances of whole groups of living things in discrete periods of time, can be accounted for only by a number of quick, convulsive, and deadly catastrophes: deluges or holocausts that would have swept all life before them, leaving a record of their passing in solid rock.

This duel between gradualism—or to give it its historical name, uniformitarianism—on the one hand and catastrophism on the other has seen many a thrust and parry. Some scientists have even sought to bridge the chasm, thrusting and parrying in both directions at once, as it were, by arguing that any event, no matter how rare or cosmic, can be construed as being in the grip of some uniformitarian steady state if it is repeated or periodic.

In the public view, the catastrophists hold all the cards; the phenomena they are free to propose—colossal mid-space collisions, stellar extinctions, lunar outpourings—are so much more remarkable than those they must confute. And for the time being they appear to be in the ascendance. Evidence of catastrophism as an explanation for at least one and maybe more of the major extinctions that punctuate the earth's geologic history is very persuasive. (The focus of modern attention is on mass extinctions in the animal kingdom; there appears to be little in the fossil record to testify to similar afflictions of the world's flora.)

Evaluating explanations of mass extinctions, or even defining what an extinction is and when it happened, is a tricky business. Even agreement that mass extinctions have taken place at all has not always been universal. "The field of extinctions is a morass," says David Raup, until recently chairman of the department of geology and now dean of science at Chicago's Field Museum. Raup has argued mass extinctions both ways over the years, finally arriving at the conclusion, now broadly shared, that they have indeed taken place.

Instances of confusion and misconception reach back to ancient times. When Herodotus, the father of history, examined limestone from the great pyramids of Egypt, he noted coin-shaped skeletons that we know now are of shelled marine animals called foraminifera. His conclusion: They must be the petrified remains of lentils, a prime foodstuff of the pyramid builders.

Some perspective is needed to gauge the truly herculean task of interpreting the fossil record. First is the enormous time span involved. Fossils recently discovered by William Schopf of the University of California at Los Angeles and his fellow paleobiologists are 3.5 billion years old. They imply that life of a simpler kind must have existed even earlier. Further, since life's beginnings, paleontologists reckon, at least 250 million different species of plants and animals must have populated the planet, yet they have discovered the fossil remains of a mere 250,000. (Estimates vary, but there are probably between four million and ten million species of plants and animals alive today.) The fossil record is, to put it mildly, incomplete.

Catastrophism

"Some gaps in the fossil record are due to sheer bad luck," says David Raup. "For example, we know that the insects have been around for at least 300 million years. But during the Cretaceous period [between ap-

Going, going... Surviving dinosaurs in the late Cretaceous period. Did the catastrophe at the Cretaceous/Tertiary Boundary cause their extinction, or were they on the way out anyway?
Charles R. Knight/Field Museum of Natural History, by permission.

proximately 135 million and 65 million years ago], their fossil record is all but absent. Conditions for their fossil preservation just happened to be virtually nonexistent for an enormous block of time."

That is the kind of problem, Raup explains, that plagues perusers of the fossil record, especially those seeking traces of terrestrial organisms, including many vertebrates. Any plant or animal that lives above sea level is in an area undergoing erosion and thus is less likely to be preserved. There are, of course, exceptions—lake deposits, overlain and lithified volcanic ash, tar pits, amber. But organisms generally have a much better chance of being preserved in the fossil record if they are marine. Their remains become trapped in layers of sediments; with changes in sea level, upthrusting, and other geological events, fossil-rich seafloor material may surface to become part of the dry land and thus more accessible to science.

Examination of just such a site led to the first real notion of mass extinctions. Toward the end of the eighteenth century, laborers digging in the soil of Montmartre in the Paris basin turned up strange fossils, which were eventually handed over to the illustrious French naturalist Georges Cuvier. A careful study by Cuvier and others revealed a succession of markedly different layers, one atop the other, of varying composition. And each layer bore a different fossil imprint, many of which had no living counterpart. One layer might be permeated with multitudes of seashells, another strewn with the bones of giant mammals, another bare of any fossils whatever.

By 1812, Cuvier had shown that fossils of many marine invertebrates, as well as of such large mammals as the Irish elk, represented creatures that no longer existed. Cuvier concluded that the abrupt changes or gaps in the sequence of geological formations and the record of life they carried must have been caused by a series of "frightful occurrences" that periodically smote the earth, altering its face and wiping out life on a large scale.

"Living things without number were swept out of existence by catastrophes," Cuvier wrote. "Those inhabiting the dry lands were engulfed by deluges; others whose home was in the waters perished when the sea bottom suddenly became dry land. Whole races were extinguished leaving mere traces of their existence, which are now difficult of recognition even by the naturalist. The evidences of those great and terrible events are everywhere clearly to be seen by anyone who knows how to read the record of the rocks."

Uniformitarianism

That the evidence "clearly" supported a catastrophic view was robustly rebutted by the early uniformitarians. In 1785, a Scottish gentleman-farmer and amateur of the sciences, James Hutton, read to the Royal Society of Edinburgh a paper that eliminated convulsive cataclysms as molders of the earth. Published in 1795 as *Theory of the Earth*, Hutton's idea was that normal geological processes such as erosion and sedimentation, proceeding for long periods of time but at rates that were currently evident could explain all the changing wrinkles in the earth's face. He thus laid the foundations of modern geology. "We find," Hutton wrote, "no vestige of a beginning—no prospect of an end.... Time is to nature endless and as nothing."

The time needed for such slow and uniform changes as Hutton proposed far exceeded the total age of the earth as deduced by the theologians of the time. Archbishop James Ussher's calculation, made in the mid-seventeenth century, that God had created the earth 4,004 years before the birth of Christ, was still widely accepted. Hutton was not the first to propose that the earth was far, far older than a few thousand years, however; Georges-Louis Leclerc, Count Buffon, had earlier extended the age to some 75,000 years. Nonetheless, Hutton was roundly attacked for being heretical.

Then in 1830 Charles Lyell, an English lawyer and geologist, published the first volume of his *Principles of Geology*, a synthesis of all the uniformitarian ideas advanced to that time. The three-volume work has been called one of the most influential books on geology ever written; it was designed to torpedo the catastrophists for all time.

Young Charles Darwin read the first volume in 1831, at the beginning of the voyage of the *Beagle*. The book had been recommended by his professor of botany at Cambridge, the Reverend John Henslow, who wrote: "Read it by all means, for it is very interesting, but do not pay any attention to it except in regard to facts, for it is altogether wild as far as theory goes." Henslow, a minister, would have taken particular umbrage at Lyell's insistence that the earth's dimples, bumps, waters, and lands were made by natural forces that were identical to those acting on the planet in his own time. Lyell had written in a now-famous passage: "No causes whatever have from the earliest time to which we can look back, to the present, ever acted but those now acting, and they have never acted with different degrees of energy from which they now exert."

Coming and going

Many paleontologists, paleobiologists, and geologists now subscribe to a kind of hybrid of the two conflicting theories—a blend of uniformitarianism and catastrophism. Such a composite theory postulates gradual changes over long periods of time by Lyell's ever-present geological and evolutionary mechanisms, but it also allows for occasional puncutations—sudden changes caused by events of a catastrophic nature. Further, the extinction of species is now an incontestable fact in life. But what constitutes an extinction? After all, as one paleontologist has said, "The fate of every species is extinction." Since the year 1600, some 100 species of birds and 40 of mammals are known to have become extinct. The world list of endangered species as of May 1, 1980 (those in danger of extinction throughout all or part of their range), stands at 705 more, from the Molokai creeper to the unarmed threespine stickleback.

"The extinction of individual species," says David Raup, "has been considered a normal part of evolution. If an evolutionary group is healthy, it's generally thought that the number of new species generated [within it] will equal the number of species gone extinct. The half-life of a species is about seven million years." But there have been times, he continues, a half dozen or so by consensus, when extinctions went really off the scale.

No particular inference should be drawn from the fact that these mass extinctions seemingly occur at the ends of geological periods; that's the way the boundaries of the geological time scale were defined in the nineteenth century—on the basis of extinctions drawn from the fossil record. Thus there are widespread extinctions at the ends of the Cambrian, Ordovician, Devonian, and Triassic periods. (See the table accompanying "Before Pangea," in this *Mosaic*.) For example, at the end of the Cambrian, 500 million years ago, nearly two-thirds of the 60 families of marine arthropods called trilobites vanished, more or less abruptly. And the close of the Triassic saw the disappearance of most of the sea-dwelling ammonoids and of many land-based reptiles and amphibians.

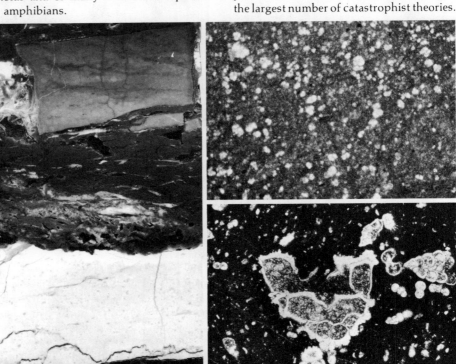

Before and after. A narrow clay layer (left, darkest band) marks the end of the Cretaceous period (white limestone) and the opening of the Tertiary (darker band above). The one millimeter foraminiferan (bottom right) in the latest Cretaceous layer was among the species that did not survive the great dying. The upper micrograph is of the earliest Tertiary rock from the same place and in the same scale.

Walter Alvarez, by permission.

The great dyings

These periods of mass extinction pale before what Raup calls the two "biggies": the ones that occurred at the boundary between the Permian and Triassic periods and at that between the Cretaceous and Tertiary.

The mass extinction at the end of the Permian, some 225 million years ago, has been called "the great dying" by Harvard paleontologist Stephen Jay Gould. It was the most severe and widespread scourging in the history of life on earth. More than half the families of marine animals, including all the surviving trilobites and all the ancient corals, were obliterated. Some 75 percent of the amphibian families disappeared, by some accounts, and more than 80 percent of the reptilian. Raup estimates that by the end of the Permian 52 percent of all families of life forms had been extinguished. And from that he extrapolates: "As many as 96 percent of the animal species then living were killed off at one swipe. That's a real holocaust."

Almost as direful—but far better known—was the mass extinction at the end of the Cretaceous period some 65 million years ago. This dying, most dramatic perhaps because it is closer to us in time, perhaps because it pushed the dinosaurs over, has stimulated the largest number of catastrophist theories.

Both land and sea creatures perished at the end of the Cretaceous. Making an exit along with the dinosaurs were the flying reptiles, the giant marine reptiles, and numerous marine invertebrates, including the ammonoids—shelled animals which resemble the nautilus and which had twice before come to the verge of extinction, once at the end of the Permian and again at the end of the Triassic. Only a few species survived among the microscopic calcareous marine plankton whose skeletons of calcium carbonate sink to the ocean floor eventually to form limestone. Significantly for the present, mammals survived.

The number of families going extinct at the end of the Cretaceous is probably somewhat lower than those wiped out at the end of the Permian, but the results are devastating enough. One estimate has 50 to 75 percent of all animal species dying out, and the evidence shows that the event was very abrupt in geological terms. Plankton disappeared from the seas within 200 years, according to Dutch geologist Jan Smit.

One testable hypothesis

Some of the catastrophist theories offered to explain extinctions, "the sort of ad hoc explanation that floods the literature, the kind of 'Just So Stories' we've been hearing for years," says David Raup, "are impossible to document, except by wishful thinking. Many plausible scenarios have been written, but they lack hard data." (See "Just so," accompanying this article.)

There is at least one recent theory of the Cretaceous-Tertiary mass extinction—and perhaps others—that does not belong in the realm of the "Just So Stories," however. It is the asteroid-collision idea, propounded most vigorously recently by Walter Alvarez, his father, Nobel laureate Luis Alvarez, and their colleagues at the Lawrence Berkeley Laboratory and the University of California at Berkeley. Briefly, it states that an Apollo-type (earth-orbit-crossing) asteroid roughly ten kilometers in diameter plunged into the earth 65 million years ago triggering the extinction.

So far, this is just one more story. But, say the Alvarez group, the giant asteroid left a chemical calling card. "That's what makes it qualitatively different," says David Raup. "The Alvarez theory is testable."

The evolution of that theory would have delighted the three princes of Serendip as well as sponsors of father-son days. Luis Alvarez is, of course, a physicist. His Nobel prize came in 1968 for the development of the hydrogen bubble chamber, among other things. "Now," he says, "I'm practicing geology without a license." His son, Walter Alvarez, is licensed. An associate professor of geology at Berkeley, Walter had been investigating a sedimentary rock formation found near Gubbio in central Italy. The rock had been submerged by the sea; later it became upthrust and was exposed to human view by the ubiquitous agency of road building.

This rock, with layers of pink and white limestone, is special. It embraces the terrible span during which so many life forms disappeared from earth: the Cretaceous-Tertiary boundary. In the rock's lower limestone stratum paleobiologists have identified the fossil remains of a multitude of minute calcareous sea creatures of the order Foraminifera. Separating the lower and upper limestones is a thin layer of reddish brown clay in which the forams, as these marine protozoa are also known, simply do not appear, and

The boundary. The Cretaceous/Tertiary boundary at Gubbio (top) and nearby at Petriccio, in Italy.
Walter Alvarez, by permission.

so the calcium carbonate content is much reduced. In the upper limestone layer, a few related forams gradually reappear.

The legend in the rock is clearly consistent with the timetable of paleohistory's most dramatic extinction. Planktonic organisms with calcium-containing skeletons, abundant in the Cretaceous (the lower limestone stratum) suddenly disappear (the calcium-devoid clay between the layers of limestones) and make a return in the Tertiary (the upper limestone).

Walter Alvarez, among others, had used the technique called magnetic stratigraphy to date the events in these and similar rock samples. The transition zone, the narrow clay boundary, could at least be located in time. But how many years did that barren, two-centimeter-thick clay layer actually represent? What, in short, was the duration of the catastrophe? That was the question Walter discussed with his father.

The two had in any event been seeking a project that might combine their interests—keeping it in the family, so to speak. The elder Alvarez, who had worked extensively both with cosmic rays and nuclear chemistry, then conceived the idea of a chemical marker by which the time span represented by the clay layer might be measured.

The key to his solution was that sediments that are slowly deposited on the ocean floor, where limestone is formed, invariably incorporate some material from space. These are primarily meteoritic materials, hundreds of tons of which fall on the planet at a known, constant rate. Meteoritic material is chemically different from earth-surface material. Among elements that are far more abundant in meteorites than on the earth's surface are members of the platinum group, including the silver-gray metal iridium.

Elements of the platinum group have an affinity for iron. They were depleted from the earth's surface when huge amounts of molten iron traveled to the planet's core during its infancy. Thus iridium is several thousand times as common in meteorites as it is on the earth's surface. Standing out from the normal incidence of crustal elements, and with its rate of deposition known, iridium can be made to serve as a cosmic calendar for rock in which it is found. "Because we knew the rate at which extraterrestrial matter rains down on the earth each year," Luis Alvarez recalls, "it occurred to me that by measuring the abundance of iridium in the sedimentary deposits we might be able to tell how long a period of time was represented by that clay layer."

The fingerprint

To measure the small quantities of iridium involved, Walter and Luis Alvarez turned to a team of nuclear chemists at Lawrence Berkeley Laboratory, Frank Asaro and Helen Michel—and a technique the chemists had helped refine: neutron activation analysis. In neutron activation analysis, material is made radioactive inside a nuclear reactor. The energy and intensity of the gamma rays emitted from such a sample, sensed by a germanium detector, then provide a signature for each radioisotope created. Working backward, the chemists can deduce the original elements present and their quantity.

When they were approached by the Alvarezes to analyze the Gubbio sediments for iridium, Asaro and Michel were not sanguine. "We'd tried to measure iridium once before as a marker for stratigraphic dating," says Asaro, "but we couldn't detect any. We looked at samples from the Gubbio boundary,

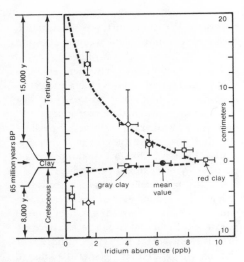

Iridium spike. Levels of iridium increase sharply when the clay layer separating Cretaceous from Tertiary limestone is reached. The evidence is of an extraterrestrial event entirely within the solar system.

After Frank Asaro/Helen Michel.

and didn't see much either. But we had just received a much larger and more sensitive gamma-ray detector. We decided to try that."

Into the Berkeley reactor went Gubbio samples from all three layers: the lower (Cretaceous) limestone, the thin layer of clay, and the upper (Tertiary) limestone. Using the new detector, Asaro and Michel analyzed the samples for iridium and some thirty other elements.

The results were totally confounding. They found iridium, all right. But it had not been deposited at a more or less constant rate. Nor was it barely detectable. Instead, says Asaro, "we found a huge increase in iridium at the clay boundary, much larger than our [pre-experimental] explanation would tolerate." The amount of iridium in the earth had jumped some thirty times the level in Cretaceous limestone. The level quickly fell back to normal in the Tertiary.

Here, then, was a totally unexpected iridium spike—a sign of a unique, major event—just at the time of the Cretaceous-Tertiary extinctions. It was a dramatically provocative finding and has changed the whole course of evolutionary debate.

No coincidence

"When we considered," says Luis Alvarez, "that this iridium peak was exactly synchronous with the Cretaceous-Tertiary boundary, it seemed not to be a coincidence." The excess iridium strongly suggested an extraterrestrial origin. Supernovas, the senior Alvarez recalls, seem to have been everybody's favorite extraterrestrial-catastrophe candidate.

Fisher is the author of "Grand Unification, An Elusive Grail," Mosaic, Volume 10, Number 5.

If a supernova occurred close enough to earth, the reasoning goes, the hard radiation emitted would either kill, or cause mutations. It would also produce a profound alteration in climate that could, according to Dale Russell, a dinosaur specialist at the National Museum of Natural Sciences of Canada, "severely tax or exterminate organisms adapted to tropical climatic conditions."

In order to have injected the amount of iridium detected at the Cretaceous-Tertiary boundary, the Alvarezes and their colleagues figured, the supernova responsible would have had to have been a mere tenth of a light-year's distance from the earth. The chance of that happening in the last hundred million years is . . . one in a billion.

"If there was only one extinction to explain," says Luis Alvarez, "such odds are acceptable, because there are about a hundred billion stars in our Milky Way." But the Berkeley group, he points out, "was looking for an explanation [that would hold] for five major extinctions in the last five hundred million years, and a supernova that close couldn't be used to explain more than one such extinction."

"This supernova theory," says Asaro, "unlikely though it was, was testable. If you make iridium in a supernova, you make other elements with it. One of them has to be plutonium 244, which is radioactive and has a half-life of more than 65 million years. So if it was laid down at the boundary along with iridium, it might still be there. We looked for it, but didn't see anything like the levels suggested by the supernova hypothesis. It is possible, however, that marine chemistry might have removed the plutonium 244."

There was another test of the supernova idea: Radioisotopes of elements forged in the intense neutron furnace of a supernova will appear in different ratios, depending on the type of star. Or, as Luis Alvarez puts it, "Cookies from different ovens have different tastes." When Asaro and Michel measured the relative abundance of two different isotopes of iridium in the boundary layer (iridium 191 and iridium 193), they found, to within 1.5 percent, that the ratio was the same as that for iridium natural to the solar system. It seemed too much of a coincidence to expect that the two unrelated stellar sources—the sun and the supernova's star—would give the same ratio. "We concluded," says Asaro, "that the iridium in the Cretaceous-Tertiary boundary layer came from an extraordinary extraterrestrial source. But the source was within the solar system."

Confirmation

Following the announcement of the Gubbio results, other findings from around the world have since driven home that conclusion. Two European geologists, the Netherlands' Jan Smit and Belgium's Jan Hertogen, reported similar anomalies in both iridium and osmium, another element far more abundant in extraterrestrial sources than on earth. They found their anomalous iridium in Cretaceous-Tertiary boundary rock found in Caravaca, in southeastern Spain. The coincidental extinction of planktonic life, they reported, "was abrupt without any previous warning in the sedimentary record." It was so abrupt that they calculate it to have occurred over a mere 200 years or so.

Then the Alvarez team found an even larger iridium spike—160 times normal—in more boundary specimens from Denmark's Stevns Klint sea cliff, 50 kilometers from Copenhagen. And an iridium anomaly has been confirmed in boundary samples from a total of five additional sites in Italy. From the other end of the world, the same anomaly was identified in a New Zealand boundary specimen secured with the cooperation of Dale Russell of the Canadian National Museum of Natural Sciences in Ottawa.

Further, the isotope ratios that label the iridium as extraterrestrial but of the solar system have held up too, confirmed by Smit and Hertogen, by Dale Russell, and by R. Ganapathy of the J. T. Baker Chemical Company in New Jersey. Ganapathy tested for iridium, osmium, gold, platinum, rhenium, ruthenium, palladium, nickel, and cobalt. Conclusion: "A close resemblance between the abundance patterns of noble metals in the [boundary] clay and in primitive meteorites."

The latest Cretaceous-Tertiary iridium finding is unique, because it is from material laid down 65 million years ago on a continental land mass, not in the sea. It therefore frees the occurrence of iridium of any connection with marine geochemistry. The sample containing the iridium comes from exposed shale near Fort Peck, Montana. It was dated by analysis of pollen remnants. The Lawrence Berkeley team is carrying out experiments to determine the size of the anomaly and its origin.

The last theory

To the Alvarezes and their team the clues were clear, but the problem was still puzzling. Sixty-five million years ago, something from within the solar system had bombarded the earth with material in such a way as to leave its chemical fingerprint all over the world and to cause the mass extinctions of an incredible diversity of creatures. But what? "I went through two theories a week for about a month and a half," says Luis Alvarez, "including novas of the sun and solar flares." Each was matched against physical and biological criteria. The event had to be something that was probable in a period of 100 million years or so. It had to account for the geochemical evidence on earth. And it had to be capable of initiating the biological extinctions."

"All but the last of our theories went down the drain," says Alvarez, "and that one was an asteroid impact. We had tried asteroids before, but all we could come up with as a result were tidal waves, which didn't seem to wipe out the dinosaurs on the Colorado plateau. But then I got the idea of a dust cloud completely surrounding the earth.

"A few years ago, at a symposium on extinctions, someone had asked what would happen if you turned out the lights, meaning dimming the sun. Most of the paleontologists there said that would result in a good match with the fossil record of the extinction. But there were no good suggestions for how to turn out the lights.

"When I went back to the asteroid theory, I calculated that one big enough to bring in the observed iridium—ten kilometers in diameter—could do the job. And the probability of such an asteroid hitting the earth, a calculation made for us by E. M. Shoemaker [of Caltech and the United States Geological Survey], was one every hundred million years—just right."

If the Alvarez team is right, what an extraordinary spectacle that must have been, an awesome apparition the like of which has fortunately never been witnessed by humans. At least once, but maybe five times in the 500-million-year span back to the Precambrian, a great white-hot chunk of rock battering its way through the atmosphere, at speeds of about 100,000 kilometers an hour, would deliver earth a hammer blow beyond imagining.

It would blast out a sizzling crater some 200 kilometers across. But beyond the shockwave ripping through the air, the violent earthquakes, the staggering tsunamis, the direct damage to life, would be the volume of matter thrown up by the impact—60 or more times the asteroid's own mass, Luis Alvarez calculates. From this pulverized debris, an immense quantity of dust would reach the stratosphere, perhaps a thousand times more than resulted from the great Krakatoa eruption of 1883.

This huge dust cloud, according to the Alvarez lights-out scenario, would have turned day into night for several years until the dust settled, with cataclysmic consequences for life on earth. The diminished sunlight would severely repress photosynthesis; most of the fragile network of food chains would be sundered. Plants would die, then herbivores would die, and then carnivores. Darkness at noon indeed.

Seabed core. The Cretaceous/Tertiary boundary shows between 103 and 105 on the centimeter scale. Immediately beneath is a late-Cretaceous carbonate zone showing evidence of burrowing by bottom-dwelling organisms; above, in the early Tertiary, are turbidite sand flows.

Deep Sea Drilling Project.

Dale Russell says that no terrestrial vertebrate larger than 25 kilograms is known to have survived the Cretaceous-Tertiary extinction. But many smaller vertebrates did survive, including the ancestral mammals. The Alvarez team believes those creatures may have been able to survive by feeding on insects, dormant seeds, and decaying vegetation.

The team is quick to point out that the notion of extinction caused by a collision with some astronomical body is hardly new. But as Frank Asaro says, "What is new is that the conclusions of this work [regarding the trigger event] are based on the scientific evidence. They are the only explanations we found that were consistent with this evidence."

It is the element of testability that makes it possible to examine the further bold surmise: Could asteroid impacts have caused other major extinctions? Luis Alvarez thinks so. After all, the chance of a collision between the earth and a ten-kilometer object is computable—one every hundred million years or so.

And this too may be proving to be testable.

Walter Alvarez recently learned of a sample of exposed, 225-million-year-old Permian-Triassic-junction limestone, carrying a clay boundary marker like that in Cretaceous-Tertiary rocks. He had been chatting with Canadian geologist James Monger, who had just returned from a trip to China. "Monger didn't know of our interest," Luis says, "and just casually mentioned to Walter a one-centimeter clay layer separating Permian from Triassic limestone. My son just about went through the ceiling." He hopes to obtain a sample of this possibly unique specimen for chemical testing. If an iridium spike is found in it, the conclusion, he thinks, will be inescapable: the same cause for both of the earth's known great dyings.

Seafloor sediments

As the stakes climb with each new find, the scientists are seeking new places to look for evidence. "Every sample [but one] of the Cretaceous-Tertiary boundary that we've looked at has been in the deep sea for at least 62 million years, and then surfaced sometime during the past 3 million years," says Luis Alvarez. "And every one has had the iridium anomaly. So the same thing has to be in Cretaceous-Tertiary sediments that are still under the sea."

Indeed, the search is on for iridium within the sediment cores wrested from far below the seafloor. But there are problems. Melvin Peterson, project manager of the famed Deep Sea Drilling Project, explains why: Except in hard-rock cores, "the rotary drill used to probe oceanic sedimentary rocks can produce distortion.... The material tends to twist and swirl, mixing sediments from different layers together and destroying the time resolution we need." If the material is soft, the dramatic, new hydraulic piston corer can return undisturbed cores. In more compact sediments, more traditional methods must be used, and results are often less predictable.

Nevertheless, there have been excellent traditional cores. One, from leg 73 of the project's research vessel *Glomar Challenger*'s crisscrossing of the world's oceans, is a core that Peterson calls "one of the best sections ever obtained anywhere through the Cretaceous-Tertiary boundary."

John LaBrecque, a research associate at Columbia University's Lamont-Doherty Observatory, was a co-chief scientist aboard the *Challenger* on leg 73, along with Kenneth J. Hsu of the Geological Institute of Zurich. "At Hole Number 524, [4,200 meters] below sea level," says LaBrecque "we drilled into the eastern flank of the Walvis Ridge, a 70-million-year-old major feature near South Africa that we now know was formed from volcanic flows. What we got was a core [315 meters] long that may be the best record of the magnetic reversals during the time of the Cretaceous-Tertiary boundary, with the possible exception of the Gubbio section."

The critical part of the core was a seven-meter-long, beige-colored stretch that spanned the late Cretaceous and early Tertiary. It captured intact the two-centimeter-thick clay boundary layer and tells the remarkable story of the Cretaceous-Tertiary extinctions in the oceans. Until this boundary section, LaBrecque says, fossils of the one-celled, surface-dwelling organisms, the foraminifera and nannoplankters, were thriving. But in the section beginning some 65 million years ago, they simply vanish. "This was a very abrupt event," says LaBrecque.

Would parts of cores apparently corresponding to the Cretaceous-Tertiary boundary event also exhibit iridium anomalies? Ken Hsu sent samples from leg 73 to Bern University for trace-element analysis. Preliminary results show that the boundary layer does indeed contain an abnormally high level of iridium, although considerably lower than the levels found in the European land samples.

There is, however, at least one hole—or its absence—in the hypothesis: If the asteroid version is correct, where is the huge crater gouged by the most recent event, 65 million years ago? No presently known crater, according to Alvarez, has the right age and size characteristics. But the impact would most probably have been in the oceans, and the

---

Just so

Modern catastrophic theories of the mass extinctions come in all shapes and sizes. Some of them rely on unlikely phenomena; some of the proponents abide in lofty isolation:

• **Dinosaurs and the ozonosphere.** Recent research warns about the destruction of the upper atmosphere's protective ozone layer by chlorine in the organic compounds used in refrigerants and as spray-can propellants. M. L. Keith, professor emeritus of geochemistry at Pennsylvania State University, suggests a similar mechanism to explain the fate of the dinosaurs. The late Cretaceous, he says, was a time of extremely violent volcanic activity. Natural chlorine from the hydrochloric acid in volcanic gases could have attacked the ozone layer repeatedly, depleting it to such an extent that it would allow the passage of high doses of lethal ultraviolet radiation from the sun. (The ozonosphere normally serves as a shield against the most damaging wavelengths of ultraviolet.) Large, bare-skinned creatures such as the dinosaurs, he suggests, would have been particularly vulnerable to the radiation. But evolving animals with newly developed feathers and fur might have survived, as would small animals sheltered by foliage or burrows. This would explain, he says, why small mammals and bottom-dwelling sea creatures survived the holocaust.

• **Dinosaur eggs and estrogens.** H. K. Erben and his colleagues at Bonn University in Germany have measured the thickness of fossil dinosaur eggshells and found that those of at least one herbivorous species became progressively thinner with time. They reason that changes in climate and habitat would have produced over crowding and intense irritability. (Crowded, intensely irritable dinosaurs hardly bear thinking about.) The German group further deduces, or intuits, that such irritability would have increased estrogen levels in the females. Result: very thin eggshells, which presumably were too fragile to permit the survival of a next generation. If you get stepped on, you go extinct.

• **Arctic Ocean spillover.** Stefan Gartner of Texas A&M University proposes an unusual villain in the Cretaceous-Tertiary disaster: the Arctic Ocean. During the late Cretaceous, he suggests, about 80 million years ago, the Arctic was a freshwater body isolated from the world's saltwater oceans; the Labrador Sea, Bering Strait, and Baffin Bay had not yet opened, so the Arctic Ocean was landlocked.

But about 65 million years ago, just at the end of the Cretaceous, a passage opened between Greenland and Norway. The result was a giant spillover of freshwater from the Arctic into the North Atlantic, and eventually over all the world's oceans, so that their surface layers became brackish. In what Gartner calls a "triple whammy," the "more or less instantaneous extinction of much of marine biota" followed, caused by the simultaneous, drastic reduction in salinity at the surface, intense oxygen depletion below the surface (because of the overlying blanket of fresh water), and chaotic disruption of the marine food chain.

And that's not all. The sudden influx of torrents of cold Arctic water would have caused a world ocean temperature drop on the order of 10 degrees Celsius. And that could have affected world climate profoundly, lowering temperatures and cutting rainfall by an estimated 57 percent. The combination of drought and cold would wipe out the prodigious vegetation of the Cretaceous everywhere but in the tropics, and the food supply of the dinosaurs would vanish. "When food or water ran out in their territory," Gartner speculates, "most of them simply died."

• **Greenhouse effect.** A different climatic explanation comes from Dewey M. McLean of Virginia Polytechnic Institute's geological sciences department. He has attributed the disappearance of large marine and land reptiles and marine plankton at the Cretaceous-Tertiary boundary to an increase in surface temperature caused by a rise in the level of carbon dioxide in the atmosphere. (This gas permits radiant energy from the sun to pass through the atmosphere, but blocks the exit of infrared radiation generated when the surface is heated. This contributes to an overall warming, called the greenhouse effect.) McLean suggests that an increase in carbon dioxide stemmed from a decline in the activity of phytoplankton, microscopic marine plants that extract carbon dioxide from the atmosphere during the process of photosynthesis.

McLean has recently concluded that the great mammalian extinctions of the late Pleistocene, clustering some 11,000 years ago, are also attributable to climate changes. If the world warmed up too fast at the end of the last ice age, he suggests, the reproductive systems of large mammals might have been unable to cope with the heat load. (He notes that even now the fertility rate of cattle falls precipitately with even modest temperature rises.) And so the mastodon, the giant ground sloth, the giant armadillo, the saber-toothed cat would have perished.

But the fate of these large Ice Age mammals, especially in North America, where perhaps 70 percent of their genera went extinct, is muddied by the presence of people. Some authorities believe that Paleolithic humans at least contributed to the extinction of these magnificent creatures.

This contribution could have been manifested in two ways: The pressure of an expanding human population could have squeezed the *Lebensraum* of the big mammals. And the early hunters might have slaughtered the beasts in great numbers, rather as the hunters of the American West did the bison. Evidence exists of large-scale prehistoric roundups

seafloor has not been mapped on a fine enough grid to find such a feature. Moreover, tectonic activity in the last 65 million years could have subducted the plate bearing the crater under an adjacent one. (See "Vulcan's Chimneys: Subduction-zone Volcanism" in this *Mosaic*.)

Yes, but...?

When juxtaposed, all the findings give massive support to the idea of some extraterrestrial impact on earth just at the time of the Cretaceous-Tertiary boundary and its mass extinctions. Many scientists are comfortable with the asteroid hypothesis, but uncomfortable with the biological scenario being offered as its aftermath. Among these are paleontologists with impeccable credentials.

Harvard's Stephen Jay Gould, for example, says that of the various aspects of the Alvarezes' life-extinguishing scenario "that don't make sense, my favorite one is the way he wants plants to survive through the dormancy of seeds, and have mammals survive by eating the seeds. It's hard to have it both ways." But with the Alvarez theory, Gould says, "the important and irrefutable evidence is the iridium. Whether or not that implies an asteroid or whatever, for the first time there is a testable theory, something you can find evidence for and look at. The Alvarez findings are very impressive. I'm all for them, and I hope they're right."

---

in which sizable numbers of animals were stampeded off cliffs to their deaths.

• **Magnetic reversals.** Every so often, say between fifty thousand and a million years from event to event, the earth's magnetic field goes through a surprising metamorphosis: It changes north-south polarity. For example, seven hundred thousand years ago (the time of the last such magnetic reversal), the magnetic north pole was somewhere in Antarctica. During these flip-flops, the field strength sinks toward zero and then builds up in the other direction (much like a sine wave).

The magnetosphere, like the ozonosphere, acts as a shield for life on earth, this hypothesis observes; it deflects potentially harmful charged particles that bombard the planet from space. Over the years, many scientists have argued that during these intervals of a lowered magnetic barrier, when the surface is relatively naked to harmful radiation, some life forms might suffer damaging genetic changes and become extinct. James D. Hays of Columbia University's Lamont-Doherty Geological Observatory has pointed out coincidences in the extinctions of eight species of the elegantly architected one-celled animals called radiolaria in the last 2.5 million years. On six separate occasions, a species has vanished immediately after a magnetic reversal. (Walter Alvarez reportedly found a 65-million-year-old reversal in the Cretaceous-Tertiary boundary.)

When is the next reversal due? Perhaps in a mere twelve hundred years, according to data obtained by Magsat, the Magnetic Field Satellite. The spacecraft confirmed a slow and steady decline in the strength of earth's magnetic field. If the rate of decline were to remain constant, a reversal in twelve hundred years is probable. (Roy Plotnick of the University of Chicago most recently cast doubt on this whole proposition: An examination of the data, he says, discloses a sequence that is just as easily explained by chance.)

• **Ring around the earth.** John A. O'Keefe of the Laboratory for Astronomy and Solar Physics at NASA's Goddard Space Flight Center has proposed a unique explanation of the mysterious conclusion of the Eocene epoch of the Tertiary period, some 34 million years ago. During the "terminal Eocene event," the mild winters of the northern latitude became frigid as temperatures plunged 20 degrees Celsius, though the summers remained mild, and many species, including small marine organisms, vanished. What led to this result, O'Keefe suggests, is that a volcano on the moon erupted, hurling billions of tons of debris toward earth. Some of the material, perhaps a billion tons of it, fell through the atmosphere to scatter in an enormous arc, called a strewn field, of the small glassy stones called tektites. (O'Keefe for years has devoted considerable effort to demonstrating a lunar origin for tektites. He has little support from his colleagues in this.)

There is a tektite field that stretches from the Caribbean halfway around the globe to the Pacific and Indian Oceans. But by far the larger mass of ejected lunar material, says O'Keefe, some 25 billion tons, formed a thin, wide ring system in the plane of the equator, giving the earth much the appearance of Saturn. Such a ring, extending from 6,000 to 19,000 kilometers above the surface, would block some 75 percent of the sun's light. Its shadow would affect principally the northern latitudes of the Northern Hemisphere during its winters, when the sun is over the Southern Hemisphere. The eerie ring, O'Keefe calculates, would not have dissipated for a million or more years, ample time to bring the Eocene to its intemperate end.

This scenario can explain why winter temperatures, but not summer ones, plummeted in the north. And there is an intriguing coincidence turned up by Billy P. Glass of the University of Delaware. His was the group that recently showed, by examination of deep-sea sediment cores, that the so-called North American tektite field actually circled half the earth. In probing the cores, he noted that the extinction of radiolarian species some 34 million years ago coincided with the sudden appearance of tektites.

• **Cometary collision.** The notion that one or more comets may have slammed into earth with catastrophic results is old. The massive extinctions of both the dinosaurs and marine organisms at the end of the Cretaceous can be explained, some say, by the impact of a comet with a mass of a trillion metric tons—about the size of Halley's comet.

The energy of so huge a body rushing through the atmosphere would have raised the air temperature by a large amount. A calculation by Harold Urey set the figure at 190 degrees Celsius in its immediate vicinity. It could even have produced local nuclear or even thermonuclear reactions. The rise in temperature would probably have killed off the large land animals.

Moreover, the comet would probably have borne significant amounts of poisonous hydrogen cyanide and methyl cyanide. (Both were detected in 1973 in Comet Kohoutek.) If the comet landed in the sea—a reasonable assumption considering that the oceans cover more than 70 percent of the earth's surface—it would have contaminated the waters with deadly cyanide, thereby dispatching large numbers of marine organisms, especially those floating near the surface. And a large injection of cometary carbon into the seas could further disrupt the ecological balance for plankton.

• **Dinosaurs' tummy troubles.** And then there is the modest proposal whimsically put forth by the English paleontologist Anthony Hallam. On account of a drastic change during the Cretaceous in the kinds of plants the huge beasts munched on, "one is led ineluctably to the conclusion that the poor dinosaurs died of constipation." •

Similar hopes—and reservations—have been expressed by the Field Museum's David Raup. "I try to keep Alvarez's geochemical work and his biological interpretations separate. Many biologists that I've spoken to are horrified by his biological scenario. And I don't really take his 'turning the lights off' with a dust cloud seriously. I also don't think it's necessary. But that doesn't affect his asteroid theory.

"If a ten-kilometer asteroid struck the earth, particularly in the oceans, my theory is that all hell would have broken loose. I don't think we're in a very good position to say exactly how the extinctions would have occurred, but they would have been significant.

There are other scientists who acknowledge the overwhelming evidence for an extraterrestrial collision, an indubitable catastrophe, 65 million years ago, but who hold out nonetheless for a more uniformitarian explanation for the biological effects. John LaBrecque is one. "I'm personally uncomfortable with the mechanics for extinction in Alvarez's theory," he says. "It's possible that an Apollo-type object did indeed strike the earth, but I think you can explain the extinctions on the basis of plate tectonics and temperature change. We do know that there were major changes in plate motion at the end of the Cretaceous. I'd be much happier with such a hypothesis than with an extraterrestrial collision to account for extinctions."

A leading exponent of a noncatastrophic Cretaceous-Tertiary extinction, geologist and evolutionist Thomas Schopf of the University of Chicago, doesn't think the Alvarez biological scenario makes sense for most creatures, or that it "applies to dinosaurs at all. Most of the dinosaurs were already extinct," he argues.

"The dinosaur extinction problem," Schopf says emphatically, "boils down to what happened to 15 or 20 species during a two-to-three-million-year interval in one area of the earth: a large seaway that extended from the Gulf of Mexico to the western margin of Alberta. Dinosaurs elsewhere in the world, as far as one can reasonably infer, all had gone extinct at earlier times. Now during this two-to-three-million-year interval, the sea level as a whole was being lowered by 100 to 150 meters. The dinosaurs didn't know what was happening. Their habitat, marginal to this ancient sea, simply disappeared gradually. So there is no magic, no catastrophe needed to explain their demise. It's simply a question of bad luck.

"As far as marine species are concerned," Schopf declares, "that has to be judged on its own merits. But I don't think you can group all these things together in one simpleminded explanation. I have no doubt that there is a geochemical anomaly of some kind. But whether it had anything to do with the biological world, and in what way, is another question. I don't think it had anything to do with the dinosaurs."

And for the even more massive Permian-Triassic extinctions when, as Schopf says, half the marine families of the world disappeared, he offers a noncatastrophic, gradualistic theory: a gradual reduction in faunal provinces, regions of the earth that can support different kinds of animals.

Over a period of many millions of years during the Permian, says Schopf, plate tectonics very slowly moved discrete continental land masses toward one another until they wedged together in the supercontinent that geologists have dubbed Pangea. But this process eliminated many faunal provinces, regions along the perimeters of the previously separate land masses. "When you do that," Schopf says, "when you create, for example, the Ural Mountains by ramming two tectonic plates together, you destroy the marine fauna that were living along both edges of an ocean. This continental movement was a phenomenon that happened on the order of centimeters per year—nothing big or sudden. But it was inexorable, and it reduced the number of faunal provinces by about one-half. And the diversity of species depends on the number of these regions."

Coupled with this slow but ineluctable crunch, Schopf says, was a lowering of the sea level, perhaps related to seafloor spreading toward the end of the Permian. This additionally reduced the size of the individual marine faunal provinces. The two effects combined, says Schopf, would account for the havoc among Permian marine populations.

There are plenty of other scientists willing to take up cudgels against catastrophic hypotheses. William Clemens, professor of paleontology at the University of California at Berkeley, thinks many gradual changes at the end of the Cretaceous contributed to the extinctions, even though "we can't rule out the banana peel on the last step, a final cataclysm that knocked the dinosaurs off." But "it's just not very convincing to say that the animals that survived the extinctions [such as the placental mammals—moles and shrews—as well as a crocodilian reptile and perhaps even the earliest primates] ran around for five years eating seeds and insects."

Cometary holdouts

Then there are catastrophists, even extraterrestrial catastrophe advocates, who hold out against the Alvarezes' asteroid. Kenneth Hsu agrees that the geochemical evidence argues for a collision—but with a trillion-ton comet or a very large meteor.

Hsu and his colleagues at the Swiss Federal Institute of Technology found major changes in the isotopic composition of carbon and oxygen within the 65-million-year-old DSDP cores. They concluded that there was indeed a Cretaceous-Tertiary horror story, with drastic changes in both ocean temperatures and ocean chemistry in the relatively short time of 50,000 years. The impact of the comet in the sea directly or indirectly caused a massive wipeout of floating marine organisms, and the oceans became desertlike for some 50,000 years.

The extinction of marine plants in turn could have diminished oxygen levels in the atmosphere. They speculate that water molecules thrown into the stratosphere by impact remained there long after any dust had settled out, contributing to a greenhouse effect and increasing temperatures on earth. In this scenario, it was the hothouse climate coupled with a reduced level of atmospheric oxygen that did in the dinosaurs.

Similarly John Wasson, head of the UCLA team that analyzed other deep-sea cores, finding at least one iridium peak, firmly believes that the impacting body had to be a comet. The proportions of extraterrestrial material he found are not typical of an asteroid, he says, unless it was metallic rather than chondritic, a possibility he discounts on astrophysical grounds. "My best guess is that it was something else entirely," Wasson says. "A very weak body—a comet—like the one that produced the great Tunguska explosion in Siberia in 1908, but about ten million times as big. It could have broken up on entering earth's gravitation field, and then fragmented again in earth's atmosphere."

There are other theories. But that's the main line. The controversies only cursorily explored here, as convoluted as the great tree of life itself, will go on for a long time, propelled by humanity's conflicting ideas about its nature and by its unquenchable desire to understand its home and its history. And then there is the sheer intellectual pleasure of discovery, which, one can be certain, will guarantee an unbroken line of inquiry against all odds. To quote Charles Darwin again, in a letter to his sister Catherine written in 1833 from the pampas of South America:

"There is nothing like geology. The pleasure of the first day's partridge shooting or the first day's hunting cannot be compared to finding a fine group of fossil bones, which tell their story of former times with almost a living tongue. . . ."

The National Science Foundation contributes to the support of research discussed in this article through its Geology, Geochemistry, and Systematic Biology Programs.

"Feedback" Produces a Theory of Ecology

The intricate involvement of living things with each other's evolution produces *coevolution* and some notions Darwin had an inkling of.

From a near-earth orbit, this planet's biosphere is a thin, patchy, undifferentiated green and brown layer sandwiched between lithosphere and atmosphere. From its appearance, this pullulating, sun-bathed membrane, ceaselessly exchanging matter and energy with its environment, might appear to be a single, globe-enveloping, tree-high and sea-deep organism with a complex metabolism.

It isn't, of course. But so integrated a view of the earth's biosphere is not all that arcane. As late as the early 20th century, many life scientists regarded biotic communities as supraorganisms of a sort—integrated entities in which component species evolved to fulfill intricate, coordinated roles comparable to those that tissues and organs serve in individual creatures. This holistic concept of life on earth gave way in later years to another: that organisms interact with their physical environment more than with each other, and that species evolve by incorporating the genes of the fittest individuals among them better to withstand the stresses that the environment imposes. Only in the last decade or so has this 19th-century, survival-of-the-individually-fittest view of evolution begun to be elaborated to encompass a modernized web-of-life outlook.

Overwintering monarchs. Monarch butterflies, overwintering in Mexico, can make so great a migration in part because of an evolved ability to exploit an insecticide produced by their favorite food plant.

George D. Lepp

Yet Charles Darwin, who with A. R. Wallace and others launched modern evolutionary theory 120 years ago, had written in his landmark *Origin of Species:* "... The structure of every organic being is related in the most essential yet often hidden manner, to that of all the other organic beings with which it comes into competition for food or residence, or from which it has to escape, or on which it preys...."

Darwinism, under the influence of 19th-century individualism, came to be construed simplistically as the "struggle for existence" and "survival of the fittest." And the latter was understood—not quite accurately—as the fittest individuals prevailing over a hostile physical environment and pitiless adversaries to pass their successful attributes on to their progeny. But Darwin had not meant that the most fit necessarily survived over the dead bodies of their fellows and their foes. Rather, out of the confusion of circum-Darwinian sources on which insights into the evolutionary process depended, what seems to have been implied was that species evolved to fit their environments in a great variety of ways. These included the forming of mutually beneficial partnerships with other kinds of organisms.

Changing views

In Darwin's day, research "apparatus" for the study of evolution consisted essentially of sharp eyes and notebooks in which to record the varieties of observed species and evidence of the forces that drove their development. Then, as the 20th century turned, Mendelian genetics began to explain the mysterious ways in which nature moves to evolve species.

And in a brilliant synthesis, in 1920, an American and an Italian mathematician separately but simultaneously announced their discovery of nature's law governing the fluctuation of populations. (See "The Strange World of Chemical Oscillations," *Mosaic*, Volume 9, Number 4.) Alfred Lotka of Johns Hopkins University analyzed the three-way, autocatalytic interdependence of populations of foxes, rabbits and grasses. Vito Volterra worked out similar equations to predict annual abundance and scarcity cycles for fish in the sea.

Lotka's theory of species interdependence and Volterra's counterpart concept dominated population biology for nearly half a century, until a new generation of investigators, equipped with computers and subtle tools of biochemical analysis, began to build on the theoretical base they had inherited. They were able to see that Lotka's foxes and rabbits were affecting not only each other's numbers, but were actually influencing each other's physical evolution—and that of the grass as well.

The new name for this old-new idea is "coevolution," a word coined in the nineteen-sixties by population biologists Paul R. Ehrlich and Peter H. Raven, both then at Stanford University. It reflects a triple marriage of evolutionary biology and population genetics with the discipline of ecology. In the decade and a half since then, coevolution has permeated the thinking and exploring of more and more life scientists. In the last 10 years, for example, unsolicited research grant proposals to the Ecology and Ecosystem Studies Programs of the National Science Foundation, where a bulk of the support for these disciplines is concentrated, have increased from the rare proposal a decade ago that might be considered coevolutionary in emphasis to between some 20 and 40 per cent of the programs' totals now.

Not that coevolution has replaced or repressed the individualistic view of fitness and survival. That viewpoint held, indeed still holds, that each plant and animal is tested mainly in a one-on-one struggle with its environment. Specific interactions with other life forms—symbiosis, predation, mimicry and the like (see the glossary accompanying this article)—are considered simply to be unspectacular special cases of the notion that species influence each other by modifying or being elements in each other's physical and biological environment.

Feedback

In emphasizing that species evolve by making an incredible variety of direct, reciprocal adjustments to each other, the new discipline of coevolution is still feeling its way and taking its own measure. Perhaps coevolution's main contribution so far to the understanding of the mechanism of evolution is the principle of positive feedback. That is, when one form of life evolves a way to defend against—or to exploit—another, the other in turn tends to evolve a countermeasure to weaken or nullify that defense—or exploit the exploiter in turn—and so perpetuate itself as a species.

"All living things have the innate tendency to increase and convert as much of their environment into themselves and their young as possible," declares ecologist David Pimentel of Cornell University.

"Competition," notes evolutionary biologist Daniel H. Janzen of the University of Pennsylvania: "results in specialization; specialization results in all the different kinds of living things that we see on the earth's surface." Further, evolutionists often cite the adage that, across geological lengths of time measured in millions of years, the impossible becomes possible, the possible probable and the probable certain. The coevolution of intricate webs of feedback characteristics between and among mutually dependent species can surely be considered a case in point.

The way of a wasp with a fig

Consider the implausible reproductive strategy of the wild tropical fig. Each unripe green fruit, a package of 100 to 1,000 small ovules that, when fertilized, will develop into seeds, contains a single tiny hole or pore in its fleshy skin, corked by scales. Along comes a minute female fig wasp laden with pollen from a distant fig tree. Lured by a chemical attractant that the fruit secretes, she squeezes through the pore's trap door and enters. Then this short-lived, symbiotic insect deposits pollen, fertilizing the ovules, and lays one wasp egg in each of many of the developing seeds before she dies.

A wee wasp larva hatches, eats the seed and grows inside it for a month or so. Then mature, wingless males emerge, cut into female-occupied seeds, copulate, and then together the males drill a tunnel through the wall of the ripening fruit. Via this escape route, the females emerge into daylight, pick up a burden of pollen and fly off to pollinate other fig trees.

"The plant pays," says Janzen, who has studied it, "50 per cent or more of its offspring (seeds) for outcrossing services by the wasp." Outbreeding offers a plant's varied genes maximum play to recombine and produce diverse offspring, the fittest of which will survive to reproduce successfully in their turn, leading to further evolution.

But that's not all. Two more details illustrate the lengths to which natural selection can go, given time and feedback: When the entering female wasp wriggles her way down that pore, the form-fitting scales that cork it strip the insect's wings off—she won't be needing them any more—and cleanse her body of bacteria-laden dirt, yeast, fungal spores and other microorganisms that might infect the seeds. Then the fruit bathes the intruding insect in a special antibiotic solution to make doubly sure the dying wasp is purged of germs. Nonetheless, these tiny pollinators carry in with them their own specialized parasites—wasp mites an order of magnitude tinier—that escape the stripping and disinfectant processes, adding another level of life to the fig-centered biotic community.

Nor is *that* all. In the tropics, dozens of birds and bats and other mammals eat figs. They do the plant the service of dispersing its fertilized seed in their dung, to take root out from under the shade of the parent tree.

This intricate, intimate fig-wasp mutualism, with overtones of wasp-mite, fig-mite, fig-bat and fig-bird symbiosis, seems like a far-fetched figment of a coevolutionist's fantasy. Nevertheless, Charles Darwin prefigured it: "... I can understand," he wrote, "how a flower and a bee might slowly become ... modified and adapted to each other in the most perfect manner by the continued preservation of all the individuals which presented slight deviations of structure mutually favorable to each other." Much of such coevolution pairs species of fauna with species of flora.

One plant's poison

Most insects live by eating plants. But most plants, thanks to elaborate chemical defenses, are inedible or toxic to most insects. Besides the substances it needs for its own growth and reproduction, a plant diverts some of its energy to the production and secretion of what appear to be secondary compounds, including insecticides, to protect it from herbivorous parasites.

Thus the coffee bean, the tea leaf and the cacao bean all contain bitter alkaloid poisons—caffeine, theophylline and theobromine respectively—to repel their insect enemies. But against each such chemical weapon, a specialized bug in turn evolves its own defense.

An insect coevolving in specific interaction with a particular plant may develop ways of detoxifying an unwholesome chemical, as by secreting an enzyme to break it down. It may even perfect metabolic machinery for turning the toxin into food. Or it may pack the poison away whole in or between cells of its body. These strategies make it possible for a single insect specialist to coexist with its plant partner in a chemical love-hate relationship. It also affects that insect's fate as prey for its own natural enemies, higher up the food scale.

Biologist Lincoln P. Brower of Amherst College in Massachusetts studies milkweed and its coevolved consort, the monarch butterfly. The lowly milkweeds are a family of shrubs, vines, herbs and weeds—plus fancy garden flowers—that all exude a milky latex. This contains a cardiac poison so deadly that South American Indians used to tip their arrows with it; milkweed latex is used to make emetics and a variety of other medicinal drugs for treating bronchitis, cancer and heart disease.

A Glossary

Close encounters of a never-ending kind among plant and animal species influence their existence and fuel their evolution. These interactions are varied, as are the pressures and processes that provoke and mold the changes whereby species cope with their environments. In coevolutionary terms, "environment" encompasses the species with which one coexists along with surrounding habitat and the physical elements.

Here, broadly defined—in logical rather than alphabetical order—are the key words by which evolutionary biologists describe these processes and interactions, as exemplified in the accompanying article:

EVOLUTION: The genetic process by which an organism "changes its spots"—develops new traits from generation to generation.

NATURAL SELECTION: The major force driving evolution toward the differential reproduction of species with sets of traits adaptive for their survival and reproduction. In their ecological interactions, such as competition or predation, plants and animals exert selective pressures on and offer opportunities to each other.

ADAPTATION: The process by which a species acquires traits that improve its chances of surviving and reproducing in its specific environment.

COEVOLUTION: The reciprocal evolutionary changes that result from two or more unrelated species exerting adaptive pressures on each other. These naturally selective interactions may involve *predation*, *competition* or *symbiosis*.

Predation: One animal (the predator) killing and feeding on another animal or plant (the prey).

Carnivory: A form of predation in which animals eat animals. (It can also take the form of *parasitism*.)

Herbivory: A form of predation in which animals eat plants. (It too can be *parasitic* in nature.)

Specialist: A predator adapted to a particular species of prey.

Generalist: A predator fitted to a variety of prey species.

Competition: Two closely associated species striving, at each other's expense, to obtain the same limited resource.

Symbiosis: Two species living in close non-competitive association, to the ecological benefit of at least one of them.

Mutualism: A symbiotic relationship of benefit to both species, enhancing their ability to survive and reproduce.

Parasitism: Symbiosis to the detriment of one partner, the host, on which the other, the parasite, feeds, or from which it otherwise extracts energy, but without killing it.

Commensalism: Symbiosis to the benefit of only one party, but inflicting no harm on the other.

CONVERGENT EVOLUTION: The process whereby differing species independently evolve similar traits as a result of living in environments with similar selective pressures.

Mimicry: A special case of convergence in which one species evolves to imitate another in shape, color, odor or other attribute in order to improve its own chances of survival by deceiving its predator.

The effect of the latex on birds, mammals and most herbivorous insects is sufficiently unpleasant that they learn to leave milkweed alone. But the same evolutionary pressures that drove the plant to perfect its chemical deterrent pushed specific monarch butterfly species to adapt for neutralizing and taking advantage of the poison.

These big, boldly patterned, orange and black insects deposit their eggs on the milkweed's leaves, which the larvae greedily gobble as soon as they hatch. Monarch larvae pack the deadly, active ingredient of milkweed latex away whole, sealed off in their body cells and tissues so that the poison does them no harm. This narrow evolutionary adaptation gives the caterpillar a monopoly of succulent, nutritious milkweed leaves, free of competition from other herbivores.

In about three weeks the larvae metamorphose into winged adults, ready to mate or to fly from the United States to southern Mexico where hordes of monarchs spend the winter. It takes a certain amount of energy to transport an insect weighing half a gram 4,000 kilometers through the air; its power pack is charged from the energy in the milkweed leaves the larva ingests, lethal latex and all.

And the latex does more for the butterfly. It arms the winged adult monarch with a weapon so formidable that birds think twice before trying to eat one a second time.

Brower has fed hungry bluejays on milkweed-fed monarch butterflies. The birds promptly vomit. The scientists determined that half the dose of chemical that could kill a jay was enough to trigger the bird's brain center for regurgitation. The monarchs don't kill the jays, but they do discourage them.

From such experiments the Amherst researcher has propounded a three-stage

More than immune. A monarch butterfly larva dines on a milkweed leaf, storing what for other herbivores is a deadly poison.

Lincoln P. Brower

"gourmand-gourmet" hypothesis: (1) A bird with no prior experience pounces on a monarch, bolts it down, throws it up and, thereafter, remembers the insect that made it sick. (2) Next time the bird spies a butterfly, it cautiously samples the flavor like a gourmet instead of gobbling like a gourmand. (3) Soon the bird knows enough to reject butterflies of the size, shape, markings and colors that distinguish the unpalatable and potentially fatal milkweed-fed monarch. Thus the insect is protected from predation by the very plant poison that it has co-evolved to tolerate. Further, birds that don't regurgitate at a small enough dose will not live to propagate *their* kind.

But there's more to the milkweed-centered 3-D lattice of coevolution: Another butterfly, viceroy by name, whose larvae feed on harmless willow leaves, have evolved the same flashy wing pattern and colors as the toxic monarchs. Brower's sadder but wiser bluejays shun these imposter butterflies too. Thus, by mimicry, viceroys gain the antipredator protection of monarchs without

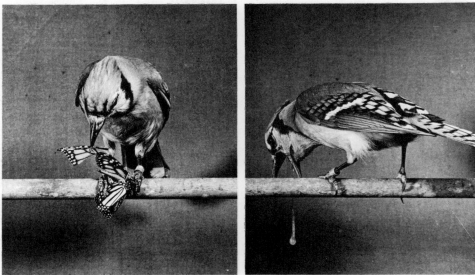

Lesson learned. A bluejay will eat a monarch once (left) and learn from the gut reaction (right) to leave monarchs alone in the future.

L. P. Brower

expending metabolic energy to compartmentalize the latex chemical out of harm's way in their tissues. Inexperienced jays, however, raised on monarchs fed on cabbage leaves, will eat such monarchs and even viceroys with relish and impunity.

But not parsnips

Nor is all the flexibility on the side of the predator. When a plant that secretes a noxious compound finds that its specialized, coevolved insect herbivore has defused it by adapting to detoxify or thrive on the repellant chemical, it may itself, in a feedback reaction, evolve the manufacture of a new poison. Cornell ecology professor Paul Feeny tested this escalation process on the voracious southern armyworm, an eat-anything caterpillar that devours the leaves of half a hundred varieties of plant, rendering all their anti-herbivore substances harmless. The generalist armyworm draws the line, though, at parsnips. This innocent-seeming vegetable secretes xanthotoxin, which is derived from another insecticidal compound, umbelliferone. The latter is produced in carrot leaves.

Feeny and his graduate students distributed 100 armyworms among five laboratory dishes. Two platoons had their standard chow supplemented by xanthotoxin—essence of parsnip—in strong and weak concentrations, and two by strong and weak doses of umbelliferone—essence of carrot. The fifth batch of caterpillars got no added chemical. Both parsnip-drugged dishes of armyworm recruits stopped developing and died; the carrot-compound groups flourished. So did the no-drug control contingent. As Feeny explains, this ability of a plant to convert umbelliferone to poisonous xanthotoxin may confer some degree of defense against herbivores already able to cope with the former insecticide.

Not only are some insects indifferent to some insecticides; some can thrive on them. Pennsylvania's Dan Janzen recently discovered a Costa Rican bean beetle that actually digests and metabolizes an organic insecticide. Its larvae feed exclusively on the seeds of a neotropical legume. Fully 13 per cent of this plant's substance consists of a potent insect-killer, canavanine. Instead of merely detoxifying this poison or filing it away in dead spaces of its body, Gerry Rosenthal of the University of Kentucky has found, the resourceful beetle larva disassembles the canavanine molecule. With the help of special enzymes, the larva converts canavanine to ammonia, a useful source of nitrogen needed by the growing beetle.

Tactics and strategies

Nor is chemical warfare the only co-adaptive anti-predator strategy available to plants. Even excepting such botanical

Neat fit. A *Heliconius* butterfly lays her eggs on a passionflower vine. Lawrence Gilbert (right) nets *Heliconius* in a Costa Rican rain forest.
L. E. Gilbert

freaks as the Venus flytrap, which can meet a predator on its own ground, survival and coevolution can take a plant in a variety of directions that one would be tempted to call ingenious, were not biologists such implacable foes of teleology—the idea that nature moves with purpose.

One striking example is played out in a row of greenhouses perched atop a flat-roofed lab building at the University of Texas campus in Austin. A throbbing beat pervades their hot, humid interior: the sound of myriad flapping butterfly wings. These gaudy, streamlined insects are *Heliconius*—passionflower butterflies. They and the plants to whose adaptations they adapt, the *Passiflora* or passionflower vine, were transplanted to Texas from the rain forests of Costa Rica by a native Texan, Lawrence E. Gilbert. He describes himself as a botanist and entomologist turned tropical ecologist.

Here is how Gilbert describes the forces and counterforces by which this plant and its specialized insect herbivore impel each other to evolve:

A newly emerged and mated female *Heliconius* butterfly cruises the tropical forest looking for just the right plant leaves on which to lay her eggs. "Right" means that specific *Passiflora* species, among at least 500, with which she has coevolved. Her initial reconnaissance involves spying out vinelike objects, then chemically testing the leaves with her forelegs. Soon this hunt-and-peck search is succeeded by more visual

orientation to the whole plant. In this, Gilbert believes, the butterfly is guided by a process of visual imprinting; females essentially remember the looks of vines they have found by experience to feel or taste "right."

The *Passiflora*, like other plants, secretes chemicals that render it inedible to most insects. Specialized herbivores, such as *Heliconius vis-a-vis* the passionflower vine, make a coevolutionary game of jumping these chemical barriers. In retaliation, Gilbert has found, certain species of the plant have developed an arsenal of nonchemical traits to fend off the butterfly's predation.

"From the *Passiflora* point of view," explains the Texas ecologist, "discovery by *Heliconius* means caterpillars munching its leaves, suppressing the plant's growth and reproduction, menacing its very survival. Since a major visible determining feature of any plant is the shape of its leaf," he points out, "we think that leaf shape is a key cue helping female butterflies to find the scattered and rare vines for which they are specialized."

How does a community of coexisting *Passiflora* species plug this chink in its armor? By evolving leaves of various shapes, mimicking those of plants not edible to *Heliconius*. So in each area of Costa Rican forest, the vines sport a wide diversity of

Adapters. *Passiflora* develop leaves of a variety of shapes to deter predatory butterflies, often adopting the shapes of alien species inhospitable to *Heliconius*.
L. E. Gilbert

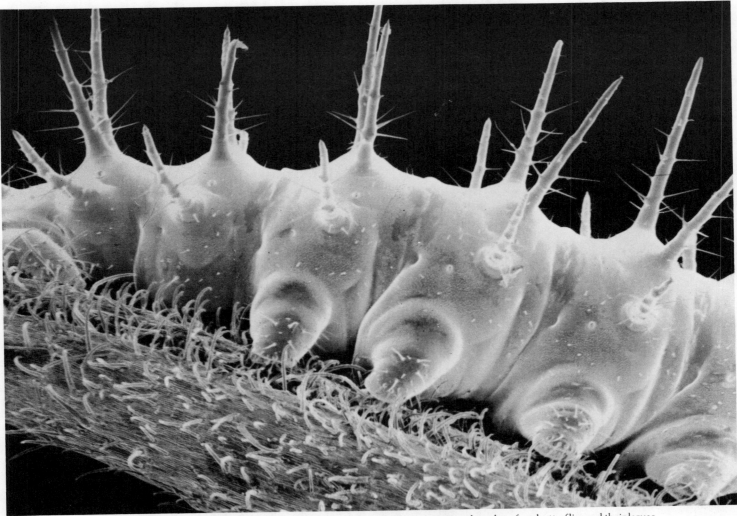

Another defense. Passionflower vines often develop leaves covered with fine, hooked hairs to entangle and confuse butterflies and their larvae.

L. E. Gilbert

leaf contours, brazenly imitating the leaves of many unrelated plants.

Finding the right plant host is only part of the female butterfly's flight mission. Not all vines of the appropriate species will offer good sites for her eggs to hatch. Some may lack the tender new shoots required by the tiny, hungry larvae. Others may harbor anti-*Heliconius* battalions of ants, wasps and other parasites and predators. *Passiflora* recruits these butterfly foes as its allies by baiting its leaves with a sugary fluid from special nectar-secreting glands. Some passionflower species have additionally evolved leaves covered with fine, hooked hairs to embrangle *Heliconius* larvae as they hatch, while deterring females from landing and laying eggs.

Experiments show that another powerful deterrent to egg-laying is a bright yellow *Heliconius* egg already in place on the leaf. The mother butterflies presumably "know," Gilbert surmises, that a newly hatched caterpillar will cannibalistically devour any still-unhatched brother or sister eggs encountered on its leaf. And he and his colleagues have discovered "at least eight *Passiflora* species that have evolved beautiful yellow mimics of *Heliconius* eggs that act to reduce the rate of egg placement upon them." No countermeasure to the fake eggs has yet been identified. Nevertheless, Gilbert regards those yellow spots as "one of the few plant traits which can be attributed unambiguously to selection exerted by a particular herbivorous insect—a clear case of coevolution."

Heliconian countermeasures to other passiflorian tactics are clearly apparent. Besides being indifferent to the toxins that keep other insects off passionflower leaves, the caterpillar of the specialized *Heliconius* happily gorges on the chemical compounds. It stores them in its body and, when as an adult insect it spreads its bedizened wings, those brightly colored pinions—as do the monarch's—flash visible warnings to hungry birds.

Besides its tolerance to the *Passiflora* toxins and its unpalatability to birds, *Heliconius* has another ally. Another tropical plant, the cucumber-like *Anguria*, welcomes *Heliconius* with open anthers, offering the foe of *Passiflora* a pollen that is a food source high in amino acids. The pollen lengthens the butterflies' life span and supplies the nitrogen, Gilbert has found, that goes to make butterfly eggs and sperm. In return, *Heliconius* pollinates *Anguria*—an indispensable reproductive service.

This triad of *Anguria*-*Heliconius*-*Passiflora* interactions is but an oversimplified tithe of the coevolved food web that Gilbert has delineated as a major subunit of neotropical forest ecology. In it he traces the ties that bind half-a-dozen plant families (representing the hundred or so actually present), their specialized herbivores and *their* specialized parasites (such as the mini-wasps that lay their eggs inside *Heliconius* eggs), plus birds, bats, bees and ants that disperse the plants' seeds.

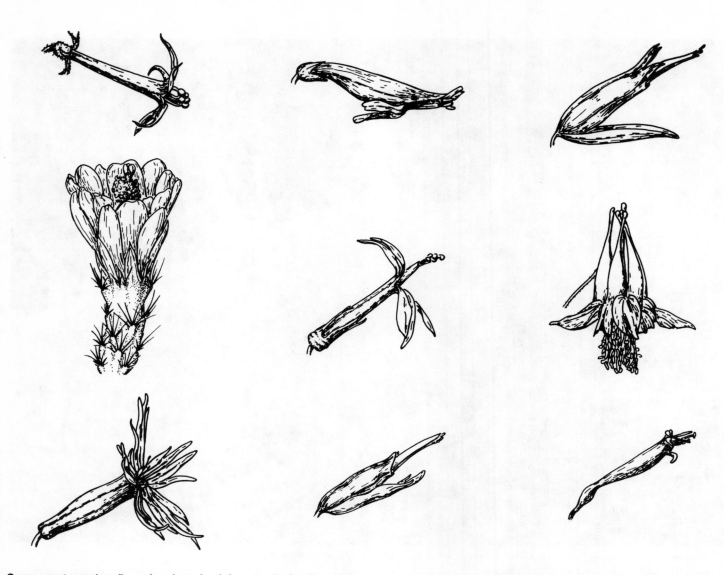

Convergent species. Examples of unrelated, hummingbird-pollinated flowers in Arizona's White Mountains—nine species from eight genera and seven families—which have developed similar shapes. From top left, clockwise: gilia, bearlip, Indian paintbrush *(C. integra)*, columbine, Arizona honeysuckle, Indian paintbrush *(C. austromontana)*, pink, hedgehog cactus flower and cardinal flower.

"It is impossible with a linear set of words," reflects the Texas scientist, "to describe such a reticulate system; we'd need a three-dimensional piece of paper."

The birds and the flowers

Obviously, then, there is more to life on earth than unending warfare between the plant and animal kingdoms. Besides battling parasites and predators, each species develops a fit with its coevolutionary opposite numbers to promote the survival and the reproduction of its kind. Since a plant is rooted to the spot where it grows, it must traffic with its enemies, as it were, paying in energy and biomass not only to attract its predators' predators, as *Passiflora* does, but to have its ovules pollinated and its seeds dispersed.

In no catalogue of survival/propagation strategies, in fact, can pollen transport be given too high a priority. It insures far more than geographic dispersal. In the endless give and take of evolution, outcrossing gives gene variants the optimum opportunity to recombine and maximizes the "fitness" of the individual—technically, the chances of the individual's genes surviving to the next generation.

To this end, hummingbirds are in some habitats a far more energetic and efficient ally than, say, bumblebees. The reward that both forms exact for disseminating a flower's sperm is nectar, a sugar solution rich in quick calories, and a flower is better served if its nectar is available to the birds but not to the bees. This is often achieved by the development of a tubular floral shape that a foraging hummingbird's bill can just enter, and of bright red petals, like airport landing lights, that migrating birds can spot in flight.

On a mile-high meadow in the White Mountains of eastern Arizona, nine flower species bloom. All are red and tubular; all but one offer nectar to hummingbirds in trade for pollination service. Among these unrelated wildflowers are Indian paintbrush, bearlip pentstemon, gilia phlox and the bright red but quite rare cardinal flower.

Two squadrons of hummingbirds compete for the floral sweets in the high meadow. The locally resident birds live year-round in the area. Rivaling the resident birds are migrant species that drop in for seasonal refueling stops early every July, on their way from as far north as Alaska, to winter camp in southern Mexico. The migrants set up and patrol compact circular territories, each containing between 301 and 1,699 flowers. This precise census was taken by evolutionary biologists James H. Brown and Astrid Kodric-Brown of the University of

Arizona at Tucson. Bird-to-bird competition is fierce, with the high-flying intruders muscling the local residents out of their territories, "competing for nectar with incredible intensity," the Browns testify.

In the sodality of hummingbird and wildflower, the Arizona scientists see a model community for studying coevolution, especially the convergence of widely differing plant species which have evolved the same red color, tubular shape and tempting nectar output in order to court the same winged pollinators.

Summer after summer the biologists reconnoiter the flowery meadows, counting flowers, sampling and analyzing their daily nectar output, observing, weighing and measuring hummingbirds, calculating their wingspread and energy expenditure *versus* caloric gain. Thus they find another remarkable convergence: that nectar secretion rates vary only some fivefold among the resident floral species. This is much less than the fortyfold variation among nonconvergent flowers pollinated by hummingbirds in Puerto Rico, and the almost thousandfold variation among totally dissimilar bumblebee-pollinated flowers that bloom together but nonconvergently in Maine.

Though the Arizona blossoms assume similar color, shape and nectar patterns to seduce the same birds, there are subtle differences among them in the way they extract transport service from the birds they lure. The Arizona researchers have discovered that some flowers are so constructed as to place most of their pollen on a bird's crown and the top of its bill; others deposit their pollen on the chin; still others at the base of the bill and front of the crown. Thus each manages to outcross selectively.

The ninth convergent species in the huge White Mountain meadow, the rare but brilliant red cardinal flower, turns out to be a freeloader. It produces no nectar at all, but grows close by the phlox and pentstemon that do, and so sneaks in on the pollination service without spending any of its own substance or energy on making sugar water.

A paradox of coevolution

Next to its pollen dispenser, a plant's best friend is its seed disperser. But Daniel Janzen discovered by chance a crucial element of evolutionary ecology: Not every animal that performs this service coevolved with its botanical beneficiary, though it may fill a niche of one that probably did.

Seed digester. A tame tapir is fed seeds experimentally in the search for a coevolved seed disperser for a Costa Rican tree.

Daniel Janzen

Made for each other. A gilia flower swaps nectar for outbreeding services provided by a male hummingbird (above) in the White Mountains of Arizona.

James Brown

In Costa Rica there is a tall tropical tree, of the mimosa family, known as the guanacaste. Its large, fleshy fruits are eaten by horses and cows, from whose dung the hardshell seeds emerge nicely scarified for easy germination. This looked like a neat textbook case of ecological mutualism, except that cows and horses were introduced into America by the Spanish conquistadors only four centuries ago, hardly enough time for them to have evolved an intimate relationship with a local tree.

The conquistadors, in fact, all but wiped out the local wild fauna, including tapirs and peccaries, which also eat the guanacaste fruit. It should be reasonable to assume—in fact had been assumed—that the domestic cattle simply took over the ecological niche and chore of dispersing the seed from these now-scarce indigenous herbivores. To verify this assumption, Janzen fed the tree's fruit to a tame tapir living on the Costa Rican ranch where he was conducting other research. The animal's dung was carefully collected and screened; it yielded few intact seeds.

Janzen's conclusion: Tapirs are not seed dispersers, but seed predators. Instead of processing guanacaste seeds in their gut for germination, they digest them.

How then account for the fact that the trees flourished long before the Spaniards supplied them with obliging cows and horses? Seeing that tapir digest the seeds, instead of passing them through his gut, led Janzen to look further back in time.

During the Pleistocene Epoch, Central America swarmed with earlier horses,

mastodons, giant sloths, nine-foot-long armadillos and other huge herbivores. Those, he reasons, more than adequately filled guanacaste's seed dispersal niche. Once they became extinct about 10,000 years ago, Janzen believes, the guanacoste probably changed its population structure and its range, with individual trees surviving where dispersal was not a problem. Then, when the conquistadors restored the horse to the habitat, the tree was able to resume its earlier distribution. (The guanacoste-horse-tapir tale underscores a point about coevolution that Janzen thinks particularly important: Not all that coexists, he points out, necessarily coevolved. And he is skeptical of many cases identified as instances of coevolution in which a thorough case has not been made for the necessary and intricate feedback patterns.)

How the grass reacts

African gazelles and wildebeeste eat grass, and African cheetahs eat wildebeeste and gazelles. The big hunting cats are reputedly the fastest beasts on feet. They have to be to catch gazelles. In coevolving, the carnivorous cat and its herbivorous prey have forced up each other's speed.

Perhaps even more intimately, though less obviously, the gazelle, the wildebeeste and the prey of both—the grass—have influenced each other reciprocally in the ecosystem where all three coexist in dynamic balance.

"The greatest and most spectacular concentration of game animals in the world live on earth's last vast unmanaged grazing system." This is how botanist Sam J. McNaughton of Syracuse University describes the Serengeti National Park in Tanzania and Kenya's adjoining Masai-Mara Game Reserve. Some 600,000 gazelle graze back and forth across this New Jersey-sized ecosystem, along with twice as many wildebeeste and smaller herds of zebra, buffalo and other ungulates.

On one May day, McNaughton observed a herd of perhaps half a million wildebeeste migrating north over a grass-covered region of the park. During the four days it took this mass of animals to eat its way through the area, it reduced the grass biomass by 84.9 per cent. During the following 28 days, the stubble put forth green shoots and replenished its substance at a rate of 2.6 grams per square meter per day, producing a short but thick green lawn.

Meanwhile, in an experimental, fenced-in, elephant-proof exclosure he had built to keep the herbivores out, the green biomass *declined* by 4.9 grams per square meter daily. Evidently, the wildebeeste graze-through, far from devastating the grassland, seemed to stimulate new growth.

How this contributes to the coevolutionary triad was strikingly borne out four days later when herds of gazelle moved in. At sites where wildebeeste had grazed, the gazelle consumed 1.05 grams of grass per square meter per day, but where the previous herd had not passed they ate nothing. And where the gazelle had passed, the grass continued to thrive.

"The association of gazelle with areas previously grazed by wildebeeste," says McNaughton, "suggests that coexistence of these two species is a consequence of coevolution that has partitioned the grassland exploitation patterns during the critical dry season. Rather than competition, there is

Picky hare. High, mature birch twigs (left in inset) grow out of reach of hares; juvenile twigs (right) are thickly covered with resinous secretions. The hare feeds selectively, carpeting the snow with resinous rejects.

F. Stuart Chapin

facilitation of energy flow into the gazelle population by the wildebeeste population...." And, by both, fertilization and stimulation of grass growth.

In harsher climes

"The whole coevolution point of view is latitude-dependent," states Texan Larry Gilbert, who pursues synecology—that branch of ecology dealing with the interactions of plants and animals—in tropical rain forests. There, he notes, coevolution proceeds most vigorously and visibly, as the greenhouselike environment promotes the proliferation, interaction and diversification of species. "It is less obvious in harsher climates," Gilbert continues, "where the options are fewer, yet these have predator-prey relationships which undoubtedly involve coevolution."

One such environment, northern Alaska's vast birch and willow taiga, is braced for an explosion. In 1980 or 1981, a population outbreak of snowshoe hares will inundate the landscape and devastate the shrubs and young trees. The furry plague erupts every 10 or 11 years, says botanist F. Stuart Chapin III of the Institute of Arctic Biology in Fairbanks. Driving through the woods nine years ago, during the last hare pandemic, Chapin recalls, he counted 10 to 20 of the teeming beasts every mile.

Snowshoe hares are the main means of support for lynx, fox and other prized fur-bearing predators of the Arctic, but are the scourge of the forest and the bane of reforestation programs. Browsing upward from the top of the snow, a hare can strip a tree bare as high as four feet above the ground.

But the birch and willow trees fight back. "The ability of a tree to tolerate browsing and still continue growing is an important part of its overall defensive 'battle plan,'" says Chapin. Its defenses include secretion of a resinous substance that complexes proteins and stymies digestive bacteria in the hare's gut. If a hare eats enough of the resin-laced bark and twigs, it starves to death.

Year by year, the plant responds to browsing by producing higher and higher concentrations of resin; year by year, the hare population declines from its 11-year high until it crashes and the cycle starts anew.

"We are basically botanists," says Chapin. "We are interested in the effect that hares have on the plants, the high level of chemical defense under pressure of the herbivores. We are studying what it costs the plants to divert so much energy from growing to manufacturing resin."

But plants cannot be understood isolated from their coevolved predators. So Chapin and his associates have conducted field experiments to correlate the hares' food preference with plant species and types of resin. Now they are feeding the resins, in a palatable mixture, to penned hares and measuring their resistance.

In other tests, the animals are offered weighted bundles of birch twigs. They eat the woody twigs but spit out or avoid the energy- and protein-rich buds and catkins on their surface, which is where the anti-herbivore resin is concentrated. A hare would rather starve, says Chapin, than consume high-resin plant parts. "This defense," he explains, "allows plants to survive the high population densities of the snowshoe hare cycles."

Underlining the bottom line

Whatever the view of earth's biosphere may be from earth-orbit, it is evident from Chapin's Alaskan forest, McNaughton's African grassland, Janzen's Costa Rican rain forest, the Browns' Arizona meadow and Gilbert's greenhouses that life on our planet is much more than a greenish-brown film palpitating between earth's crust and its atmosphere. The seamless continuum of life forms more than a web or network or mosaic. It's more like a three-dimensional lattice, with mass and energy, time and space, strong and weak forces each occupying a dimension in the constantly coevolving complexity of biotic communities.

Just as Einstein sought to comprehend all of physics in a single "unified field theory," so the evolution of all life from viruses to great whales to giant redwood trees to humans can be comprehended as an endless interplay of matter and energy pushing from the first primordial quivering molecules toward diverse, specialized, naturally selected forms.

"Humankind has not escaped coevolution," declares Larry Gilbert. "We have been able to dominate nature temporarily," he observes, "clearing forests, changing the environment, converting natural plant productivity into a few seasonal crops. In the long run—and we are not going to know how long until it is too late—our own survival really requires that these broad natural systems continue to be operative." In other words, those of a current bumper sticker: Nature bats last. •

The National Science Foundation contributes to the research discussed in this article principally through its Ecology and Ecosystem Studies Programs.

THE SIGNIFICANCE OF FLIGHTLESS BIRDS

by Edward Edelson

Flightless birds—ratites—have become a testing ground for ways to establish evolutionary relationships.

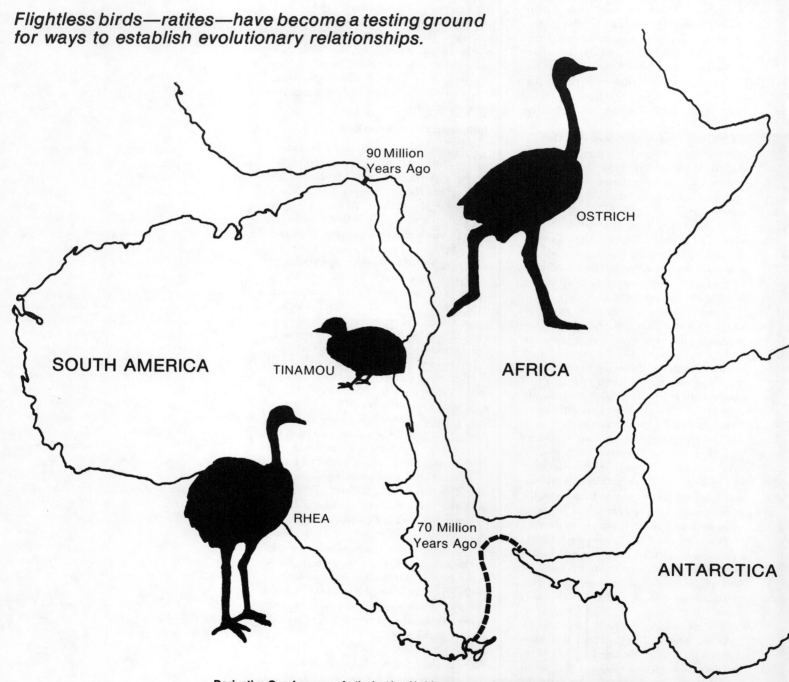

Derivative Gondwanans. As the last land bridges among the segments of Gondwanaland were severed, the ancestors of today's ratites were isolated on drifting continents, each to take its own evolutionary course.

Dates from Joel Cracraft (1972); L. R. Maxson, Vincent Sarich, Allan Wilson (1975).

Ever since they were discovered, the flightless birds—the ratites—have fascinated ornithologists and other scientists. Visually, these birds are spectacular: the ostrich of Africa, at eight feet tall and two-toed, the world's largest living bird (but nonetheless smaller than two extinct groups of ratites, the moas and the elephant birds); the two three-toed, ostrich-like South American rheas; the emu of the Australian plains; the aggressive, forest-dwelling cassowary of northern Australia and New Guinea with its two-inch, razor sharp claw; the three species of kiwi (or apteryx) of New Zealand—nocturnal birds whose five-inch egg weighs up to a quarter of the bird's body weight.

But another reason for fascination is the series of scientific questions presented by the existence of the ratites. For well over a century, the debates have gone on: Are the ratites living fossils, descendants of ancient species that never evolved flight? Or did they lose their wings because there was no selection in their environment against individuals with reduced powers of flight? Did these birds, scattered across more than half the planet, have a common ancestor and somehow disperse despite their lack of useful wings? Or did they develop from different stocks, growing to look alike through the evolutionary phenomenon called convergence, in which unlike species evolve common characteristics in response to similar spectra of environmental conditions?

Are the ratites related to the tinamous, the ground-dwelling Latin American birds that share many ratite characteristics but still can fly? And finally, if the ratites did have a common ancestor, what are their interrelationships?

Today, the common ancestry of the ratites has been established with a high degree of certainty. But, with a twist that is worthy of a Hitchcock movie, the last question has become the focus of the kind of keen controversy that often develops at the cutting edge of a science. ("It's almost the flailing edge of this science at the moment," comments one observer.)

The attempt to construct a phylogenetic tree that would establish the relationships among the ratites has become a major testing ground for an emerging science, with lively arguments about both which data to use for such an effort and which mathematical techniques are best for analysis of the body of data and its representation in a phylogenetic tree or branching diagram.

The debate about the phylogenetic branching order of the ratites thus has greater significance than might ordinarily be attached to a discussion of whether the ostrich is related more closely to the kiwi or to the cassowary. It is a gripping case history of the demanding effort being made to change the process of species classification from a subjective, ill-defined discipline to a modern, quantitative science. "This is one of the first good cases where different kinds of data can be applied to the same group," says Walter M. Fitch of the University of Wisconsin about the ratites. "That's a plus because it's an extremely interesting group."

Those interested in permutations and combinations should know that there are at least three different techniques being used to gather data about the relationships among ratites: two that examine molecules from the birds and the third, a modern version of the classical method of studying morphological differences. Additionally, several different methods can be used on the different sets of data to construct phylogenetic trees. As a result, despite a plethora of often conflicting conclusions, each proposed phylogenetic tree (and its supporting data) is thus assured the kind of close scrutiny that, ultimately, helps hone a research strategy to perfection.

Clearing the underbrush

To get to where they are, the scientists studying phylogeny via the ratites first had to clear away the underbrush by answering some basic questions. It is now generally agreed that the ratites are not primitive birds left over from an earlier evolutionary stage. Instead, they appear to be highly evolved descendants of an ancestor that had the ability to fly.

Although the ratites share many anatomical characteristics, there are some notable differences. The rhea and the ostrich have feathers that have one central shaft, for instance, while the feathers of the emu, kiwi and cassowary have two: a major one, called a rachis and a secondary "aftershaft." And, while most ratites are large birds, the kiwi is small. Further, and most obviously, the ratites are not geographic neighbors.

To those who once held to the belief that the ratites had to be an example of convergent evolution—the development of like characteristics by divergent species subjected to similar evolutionary opportunities—the biogeographic question was the clincher: How could nonflying birds spread across the vast oceanic distances between continents, virtually in all directions?

Anatomical study of the ratites, however, finally convinced scientists of the flightless birds' common ancestry. (That the ratites must have had a common ancestry, biogeography notwithstanding, had been concluded as long ago as 1867, by Thomas Henry Huxley, from comparison of physical features. And parasitologists, who consider the fact that ratites share parasites that do not plague other birds as clear evidence of common ancestry, have argued it for many years; parasites and their hosts are an example of coevolution (see "'Feedback' Produces a Theory of Ecology," *Mosaic*, Volume 10, Number 6).) But it took geology, and the evolution of the theory of plate tectonics (see "What Drives the Earth's

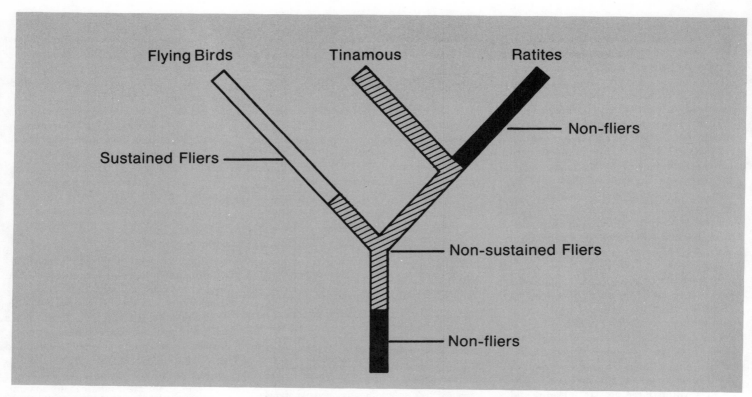

Ratite relationships. The higher the temperature at which a hybrid duplex DNA strand dissociates, the closer the phylogenetic relationship between the two species. With a natural emu duplex for a standard, Charles Sibley's molecular technique (below) makes the cassowary the emu's closest relative among the ratites and the rhea and ostrich most distant. A conjectural phylogenetic model for the origin of flight (above) suggests that the loss of all flying ability among ratites occurred only once, in a common ancestor.

Charles Sibley, Jon Ahlquist; E. M. Prager et al, Journal of Molecular Evolution 8, 283-294, (1976), by permission.

Plates," *Mosaic*, Volume 10, Number 5) to solve the biogeographic riddle: The birds had merely to stand still while the protocontinent called Gondwanaland broke into raftlike fragments that drifted apart over tens of millions of years, carrying all kinds of life forms that shared common ancestries along with them.

With that question answered, the question of phylogeny—how the ratites relate to their common ancestor and to each other—came into sharper focus. And getting a correct answer to this question is surprisingly important, since the methods used to develop an accurate phylogenetic tree for the ratites can also be used to place the ratites in their proper relationship to other birds, to develop an overall phylogenetic tree for all the orders of birds, which has not yet been done to everyone's satisfaction and, indeed, to perfect techniques for the development of quantitative phylogenies in general.

The approaches

There are three major efforts to gather data on the ratites, all three being carried out in the wider context of a study of all birds. At Yale University, Charles G. Sibley and Jon E. Ahlquist are studying the quantitative differences between the genetic material of the various ratites. Allan C. Wilson of the University of California at Berkeley has made a similar biomolecular study of one

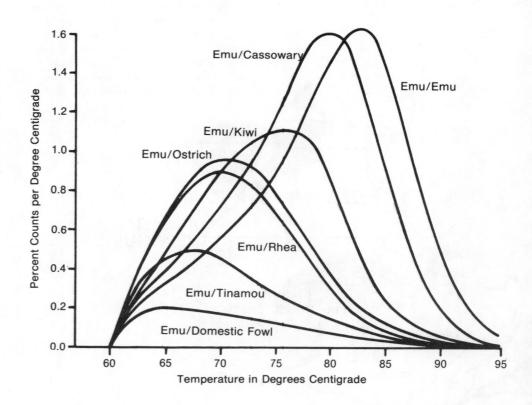

Mr. Edelson is preparing a Mosaic *article on artificial intelligence.*

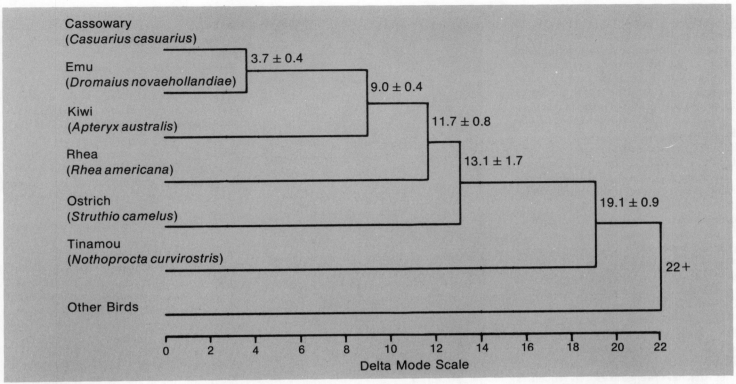

blood protein, transferrin, from the different ratite species. And at the University of Illinois, ornithologist Joel Cracraft, taking a comparative approach to the study of morphological characteristics shared by different avian species, is employing a method designed to bring more testable procedures into a process long ruled by subjective judgments.

The two molecular techniques are based on the same principle: that evolution goes on not only among physical features but also in molecules (see "Molecular Evolution; A Quantifiable Contribution," *Mosaic*, Volume 10, Number 2). The DNA molecule in which genetic information is coded is hypothesized to experience mutations at a measurable rate. The proteins produced from the

Phylogenetic trees. Results of three different ways of determining the relationships among the ratites: through nucleotide replacement in DNA (above), amino acid substitution in a protein (below, left) and modern comparative morphology (below, right).

Charles Sibley, Jon Ahlquist; Allan Wilson; Joel Cracraft, all by permission.

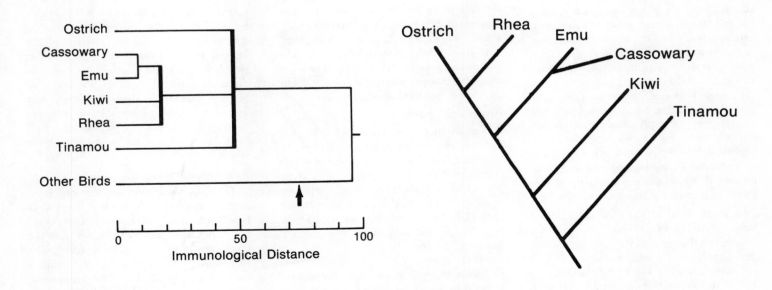

DNA instructions also show the change. By comparing either DNA or proteins from two species, an investigator can determine the extent of the difference between them and thus how far, if not when, the two species have evolved from their common ancestor. By comparing DNA and proteins from several species, data for a phylogenetic tree can be gathered.

Sibley's method is to look directly at the DNA. If purified DNA is dissolved and heated, the two strands of the DNA molecule can be separated. If the solution is then allowed to cool, the strands will reunite as complementary bases on each strand, "recognize" each other and join together. Sibley puts single strands of DNA from different species into the same solution and measures the extent to which they join. The closer the match, the closer the genetic relationship between the two species.

It is not nearly as simple as it sounds. Early work gave confusing results which were not cleared up until the discovery that the genetic material of higher species includes long stretches of repetitive DNA sequences whose function is still nuclear. Sibley and Ahlquist now fractionate the DNA to remove the repetitive sequences, leaving behind the single-copy sequences of DNA that give meaningful results. The single-copy DNA is labeled with radioactive iodine and added to a solution of nonradioactive DNA from another species. (Most of the analysis is now done automatically, by an ingenious machine designed by the Yale investigators.)

The mixture first is incubated to allow the DNA from the two different species to form hybrid duplex molecules, composed of one strand from each of the two species. Then the mixture is heated again in a series of carefully controlled steps. The temperature is raised a degree or two, and the DNA which has dissociated is removed for measurement; the amount of radioactivity in the sample indicates the degree of dissociation. By comparing the difference between the temperature required to dissociate the duplex DNA of a single species used as a control and the temperature that dissociates a hybrid duplex, Sibley and Ahlquist get a numerical reading on the similarity of the DNA strands of the two species.

A great virtue of the technique, Sibley says, is that it looks at the entire genome of a species, all the genes that count. Individual genes are known to evolve at vastly different rates; there is a 400-fold difference between the slowest-changing and fastest-changing genes, Sibley explains. By studying the whole genome, his method gives an average, overall, quantified rate of change,

Amino acid substitution. Ellen Prager: The chachalaca is not a galliform....
University of California, Berkeley.

from which he derives estimates of genetic distance between species.

Sibley describes the relationship as rather straightforward: a difference of one degree Celsius in melting point equaling approximately a difference of one percent in DNA pairing. The results are expressed in a mathematical matrix of what Sibley calls "delta modes," each delta mode expressing the genetic difference between two species. Though his colleagues are far from unanimous in agreement, Sibley maintains that the analysis of such a matrix to produce a phylogenetic tree is relatively straightforward. "The readout we get out of the computer requires very little interpretation," he declares. "The data speak for themselves."

Protein comparisons

A major biochemical role for DNA is the synthesis of proteins, and nucleotide replacement in DNA should be reflected as amino acid substitution in a protein. So the comparison of proteins from species to species, as Allan Wilson, working with Ellen Prager at the University of California at Berkeley, is doing, should also reveal phylogenetic links and distances.

Wilson's method is to purify the chosen protein—transferrin in the case of the ratites—inject it into a rabbit and harvest the antibodies produced by the rabbit. Those antibodies then are added to a mixture containing transferrin from another species and a substance called complement, which helps the antibody to join with the protein. By measuring the amount of free complement left after the antibody-protein reaction has occurred, Wilson and Prager get a measure of the differences between the two proteins.

While the method measures mutations in only a single protein, Wilson points out, "this protein is a large one, containing 700 amino acids. The number of substitutions by which the birds under consideration differ from each other is...about 100. So we really are looking at a large number of traits at once."

The method is not quite precise, Wilson acknowledges; it detects no more than about 85 percent of the changes between proteins of two closely related species. The advantage of the method, he says, is that the validity of the tree obtained with one protein can be tested by investigating additional proteins.

The technique proved itself, says Wilson, with work on a Mexican bird, the chachalaca, which had been classified as one of the galliforms, an order including such game birds as the quail and the pheasant. The study of one protein, lysozyme, indicated that the chachalaca was distant from the galliforms, a result so startling that Wilson decided to check it further. "We looked at five different proteins. All show that the chachalaca forms a group as distinctive from gallinaceous birds as ducks are." Wilson reports.

Nevertheless, Sibley disagrees. He says that his analysis of DNA shows that the chachalaca is indeed a galliform—although he agrees with Wilson that it is distant from the other gallinaceous birds.

Wilson says, nonetheless, that his conclusions about the ratites are not very different from those of Sibley. Both have the tinamous as the closest relatives of the ratites, and both have the ostrich rather closely related to the kiwi and the rhea. The differences are reconcilable, Wilson says, because imprecision that still exists in any available molecular method leaves room for changes in the resulting phylogenetic tree; "the two molecular methods agree on the whole."

A new morphology

The Wilson and Sibley constructions, however, in addition to differing in some ways from each other, differ in apparently small but nonetheless significant ways from the phylogenetic tree drawn by Cracraft. The Illinois ornithologist is using a technique of comparing morphological differences—variations in the shapes of bones and other physical traits—that is little more than a decade old in North America. The historic method of assigning relationships between species by looking at physical traits was once hopelessly subjective, Cracraft notes. But help for the anatomists, he says, came with the development some 25 years ago of a more sophisticated technique, called the cladistic method, often credited to a German entomologist named Willi Hennig.

The essence of the cladistic method, Cracraft says, is that the anatomist looks for derived characteristics the group under study might share rather than for primitive characteristics—for features that are evolutionary novelties rather than those that many species have in common. To give a most basic example, a primitive feature of birds is the presence of feathers; a derived characteristic would be their absence or a modification of the feather shared with one group of birds but not another.

There are two ways to do such an analysis, Cracraft explains. One is "outgroup comparison," a study of how a group that is under study compares with a presumably closely related group. For horses, outgroup comparison could focus on the foot structure of the order of ungulates that includes the horse (perissodactyls) or of mammals in general. The modern horse has just one toe, and outgroup comparison indicates that it descended from primitive ancestors having five toes. A phylogenetic tree might be constructed by studying the character transformations presented by many such morphological features.

A second method is to study ontogenetic transformation, the development of an individual from fetus to adult. In the flounder, for example, the eyes are on either side of the head at first, but migrate until both are on the same side of the head. This ontogenetic transformation is evidence that the primitive ancestor of the flounder had symmetrically placed eyes and that asymmetry is a derived characteristic.

The important point is that the cladistic method can make comparative morphology more rigorous and testable. The further importance of the ratites to Cracraft is that his study of them was the first application in birds of Hennig's methods.

Cracraft has found evolutionary trends in a number of morphological characteristics of the ratites. For example, studying the tibiotarsus, one of the bones in the lower leg, he finds that the base of the cnemial crest, a feature at the upper end, is broad in the tinamous, narrower in the cassowaries and the emus and extremely narrow in the rheas and ostriches. As do Wilson and Sibley, Cracraft believes the tinamous to be the closest relatives of the ratites. He concludes from his study that the ostrich is closest to the rhea among the ratites—that they are more closely related than either Wilson or Sibley would agree. Further, Cracraft and Sibley both have the ostrich and rhea more closely related to each other, so far, than does Wilson.

Which analysis?

The differences are important because they involve a critical issue in the discipline: analysis of the data. Drawing up a table of the differences, molecular or anatomical, between species is only the first step in constructing a phylogenetic tree. The next step is to apply to the data one method or another of comparative analysis to construct the tree.

Several methods of analysis can be applied to the different sets of data. One method, developed by Fitch at the University of Wisconsin, can be used on the DNA hybridization data of Sibley or the protein differences of Wilson. It starts with a matrix that lists, in tabular form, the differences between species. Constructing a phylogenetic tree can be regarded as building another matrix, read directly from the tree, which also gives the distances between species. The basic idea of this analysis is to minimize the difference between the first matrix and the second.

A different technique has been used by James S. Farris of the State University of New York at Stony Brook. Using the same table of differences between characteristics Farris's system produces a phylogenetic tree that meets what are called parsimony criteria by having branches that are as short as possible.

That the application of one system or another is far from straightforward at this stage, however, is illustrated by what happened when Sibley distributed some of his unpublished data to other investigators for comment. Farris, at the request of Cracraft, did an analysis of the Sibley data, using the method developed by Fitch. Farris and Cracraft then announced their results: Sibley's data fit Cracraft's tree better than they did Sibley's.

Sibley is unable to understand how a rigorous analysis could have produced such a result. His own subsequent analysis, he says, employing both Fitch's and Farris's techniques as well as others, resolves the distance between the emu and cassowary as well as quantifies distances from the nearby kiwi and distant tinamou. He has been unable to examine Farris's procedure, he says, but his own analyses produce not only distances between species but quantitative measures of the reliability of the distance estimates as well.

To compound the difficulties, Wilson argues further that the Fitch type of analysis "doesn't give a unique solution. You try a whole bunch of trees and estimate the goodness of the fit. You get several solutions that are almost equivalent. Our tree is a blend of the best ones."

The reason these points are important, Wilson says, is that "we are seeing a big revolution in the way people are looking at evolutionary biology. Biology is being unified in a way it never was before, and taxonomists now have to know molecular biology, as do paleontologists."

For the first time, Wilson continues, ornithologists can talk about constructing a quantitative phylogenetic tree for all the 27 orders of birds. "Despite more than a century of research," he says, "morphologists did not succeed in producing such a tree. But the main outlines of the avian family tree are now evident...." Such a biochemical tree for birds was included in an article Wilson and Ellen Prager have in the proceedings of the 1978 International Ornithological Congress.

The controversies are probably just beginning as systematic biology enters a new era of numerical measurement and analysis. "We're going to fight about methodology for a long time," Fitch says. "The truth is that our methods are just not good enough right now."

As a consequence, different actors in this drama can and do point out the shortcomings of the various approaches. Critics of Sibley's DNA hybridization technique contend that, by looking at the entire genome of a species—the method's strong point, according to Sibley—the approach smooths out the distinctive differences that could be important. Cracraft's technique draws the same criticism: "Morphology is the consequence of many genes acting on structure," Fitch says. "So the morphological data could be subject to a smoothing caveat similar to that raised against Sibley's data." Wilson himself points out the imprecision of his own data. And, against both protein and morphological analysis is raised the epigram: the farther from the gene, the farther from the truth.

The debate about ratite phylogeny thus can be seen as the forerunner of future controversies that will arise as the new methods spread through systematic biology and produce new information. "We're going to have information coming out of our ears," says Fitch. "It will be marvelous, tantalizing information. The data are going to tell us a lot. But we don't know what." •

The National Science Foundation contributes to the support of the research discussed in this article through its Systematic Biology Program.

GLOSSARY

adaptation. The evolutionary process that better fits members of a species to survive and reproduce.

anagenesis. A process that causes organisms to evolve until they gain greater control over their environment, and then acts to preserve these improvements by slowing the evolutionary process.

autotrophs. Microorganisms capable of synthesizing their requisite organic compounds.

Cambrian period. The geologic period that extends from about 600 million years ago to about 500 million years ago.

catastrophism. A school of thought that attributes events in the history of the earth and its creatures, including such occurences as mass biological extinctions, to catastrophic events such as mid-space collisions or nearby supernovas.

cladistic method. A method employed in comparative morphology in which the researcher, rather than studying all characteristics of a species, studies only those characteristics that are evolutionary novelties.

coevolution. Reflecting a triple marriage of evolutionary biology, population genetics, and ecology, coevolution involves the reciprocal evolutionary changes that take place in two or more unrelated species as a result of the pressures they have exerted on each other.

comparative morphology. The field of study that seeks to establish genealogical relationships among both existing and ancient species on the basis of anatomical similarities and differences.

convergent evolution. The process whereby differing species independently evolve similar traits as a result of living in similar environments.

Cretaceous period. The geologic period that extends from about 135 million years ago to the beginning of the Tertiary period, about 65 million years ago.

eucaryotes. Cells with nuclei. Eucaryotic organisms occupy the top rungs of the evolutionary ladder, including 99 percent of the organisms now in existence.

evolution. The development of species of plants and animals from earlier forms, by the transmission of genetic changes from one generation to the next by means of natural selection.

extinction. A dying out, annihilation, or coming to an end, as of a species of plants or animals.

fossil. The remains, traces, or recognizable impressions of an organism, preserved from some previous geologic period.

genome. The sum of all the genes in an organism.

heterotrophs. Believed to be among the earliest of life forms, these are microorganisms that are biochemically incapable of synthesizing all requisite organic compounds.

hominid. Any of a family of two-legged primates, including all humans.

hominoid. Any of a superfamily of primates, including the humanlike apes and all forms of humans.

metazoans. Multicelled animals.

microbiotas. Assemblages of primitive, microscopic life forms.

molecular anthropology. The discipline that seeks to establish genealogical relationships among species through the comparison of the species' DNA molecules and proteins.

morphology. That branch of biology dealing with the form and structure of animals and plants.

mutation. A change in genetic material which may occur on the chromosomal or molecular level.

mutualism. A form of symbiosis that is of benefit to both species in the relationship.

natural selection. The major force in evolution, natural selection enables species to develop traits that best adapt the species for survival and reproduction. According to this mechanism, individuals that bear traits giving them the competitive edge over other individuals in the species are more likely to survive and, therefore, are more likely to reproduce and pass these traits on to future generations.

neo-Darwinism. The traditional theory of evolution, this is a synthesis of Gregor Mendel's findings in genetics and Darwin's theory of natural selection. It states that natural selection, which favors traits that give their possessors a competitive edge, causes species to evolve gradually, one from another. *Also called* the modern synthesis; gradualism.

neutralist. In the field of molecular evolution, a neutralist is one who regards most evolutionary change as a result of random mutation.

ontogenic transformation. The development of an individual from fetus to adult.

paleobiology. The study of ancient life forms, associated with paleontology.

paleontology. The branch of geology that studies prehistoric forms of life through the examination of plant and animal fossils.

parasitism. A form of symbiosis in which one of the involved species is harmed, but not killed, by the species that benefits from the relationship.

Permian period. The geologic period that extends from about 280 million years ago to the beginning of the Triassic period, about 225 million years ago.

phenotype. The observable properties of an individual, resulting from the combined influences of the individual's genetic constitution and the effects of environmental factors.

phylogeny. The evolutionary history of a plant or animal.

positive feedback. The cyclical evolutionary mechanism by which one form of life evolves a way to defend against or exploit another, and the other species evolves a countermeasure, thereby enabling itself to survive.

Precambrian period. The geologic period that extends from the formation of the earth, about 4,600 million years ago, to the beginning of the Cambrian period, about 600 million years ago.

procaryotes. Very simple forms of life, procaryotes have an outer membrane but not an encapsulated nucleus.

proteins. One of the fundamental building substances of living organisms, proteins are made up of long chains of amino acids. The properties of each protein are determined by the sequence of the amino acids.

protozoans. Single-celled animals.

punctuated equilibrium. A theory of evolution stating that species experience periods of evolutionary stability that are occasionally interrupted, or punctuated, by major evolutionary changes. This theory may be contrasted with that of neo-Darwinism, which proposes that evolutionary change is constant and gradual.

ratite. A bird belonging to the order of flightless birds, *Ratitae*. The ratites include the ostrich of Africa, the emu of Australia, the South American rheas, and others.

selectionist. In the field of molecular evolution, a selectionist contends that evolutionary change occurs because it is adaptive: it is a result of natural selection.

speciation. The processes through which species diversify and multiply.

stromatolites. Rocks with stratified formations, suggesting an algal origin.

symbiosis. A form of species interaction in which two species live in close, non-competitive association, to the benefit of at least one of them. *See* mutualism; parasitism.

synecology. The branch of ecology that deals with the interactions between one species of organisms and another.

uniformitarianism. A school of thought that attributes events in the history of the earth and its creatures, including mass biological extinctions, to gradual geological and evolutionary changes. *Also called:* gradualism.

INDEX

Adaptation, 5-7, 43
Ahlquist, Jon E., 54, 56
Alvarez, Luis, 35-37, 39-40
Alvarez, Walter, 35-37, 39-40
Amino acids, 23-24, 25, 26, 28, 30, 31
Amino acid sequencing, 26
Anagenesis, 27-28
Archean life, 18
Armyworm, coevolution and, 45
Asaro, Frank, 35-36
Atmosphere, 12, 14, 15, 19, 20, 38, 39, 40
Autotrophs, 14
Ayala, Francisco, 5

Barghoorn, Elso S., 12, 14-17, 18, 19
Beck Spring formation, 19
Beetle, bean, coevolution and, 45
Biogeology, 12
Bitter Springs formation, 19
Bluejays, coevolution and, 43-44
Brower, Lincoln P., 43-45
Bulawayan formation, 18

Cambrian period, 12, 13, 14, 15, 34
Carbon-12 dating, 15, 16, 17
Carson, Hampton, 6-7
Cassowary, 53, 55, 57
Catastrophism, 32-33, 34-40
Chapin III, F. Stuart, 51
Cheetahs, coevolution and, 50
Cladistic method, 57
Cloud, Preston, 12, 13, 14, 15, 18-19
Coevolution, 41-51. See individual listings of plants and animals.
Comparative morphology, 22
  as compared to molecular evolution, 23
  as evidence of phylogeny, 23
  cladistic method, 57
  of ratites, 53, 55, 56-57
Convergence, 43, 53
Cracraft, Joel, 55, 56-57
Cuvier, Georges, 33
Cyanophytic organisms, 13

Darwin, Charles, 2, 7, 16, 31, 32, 34, 40, 42, 43.
Darwinism, 23, 30, 42. See also Neo-Darwinism.
Developmental biology, 7-9
DNA, 23-24, 27, 31
DNA hybridization, 26, 57
Drosophila, 6-7
Dugongs, evolution of, 4

Earth
  age of, 17
  atmosphere of, 12, 14, 15, 19, 20, 38, 39, 40
Ehrlich, Paul R., 42
Eldredge, Niles, 2-3
Electrophoresis, 26
Embryonic development, 9
Empiricists, 30, 31
Emu, 53, 55, 57
Environment, effects on evolution, 7, 19-20. See also Coevolution.
Eucaryotes, 14, 16-17, 19, 20
Evolution
  causes of, 5-7
  origin of, 11-20
  pace of, 2-5, 7-9, 21, 24-31
  theories of, 2-9

Evolutionary clock, 24-26, 28-30, 31
Extinction. See Mass extinction.

Farris, James S., 57
Fig Tree formation, 16, 17, 19
Fitch, Walter M., 24, 30-31, 53, 57
Fossils,
  dating as evidence of species divergence, 22
  of Cambrian period, 12, 14, 15
  of Precambrian period, 12, 13, 14-15, 16-20
  preservation of, 33
  sites of, 13, 14, 15, 16, 18, 19
  spheroidal, 16-17, 18
Founder-effect hypothesis, 7

Gartner, Stefan, 38
Gazelles, coevolution and, 50-51
Genomes, 56, 57
Gilbert, Lawrence E., 45-48, 51
Goldschmidt, Richard, 4
Goodman, Morris, 24, 26-28, 30
Gould, Stephen Jay, 2-3, 5, 7, 34, 39
"Gourmand-gourmet" hypothesis, 44
Gradualism
  as theory of evolution, 2-4, 5, 7-8, 9
  as theory of geology, 33-34
  as theory of mass extinction, 32, 33, 40
Great dyings. See Mass extinction.
Gruner, John W., 14
Guanacaste, coevolution and, 49-50
Gunflint Iron formation, 15, 17, 18

Hares, coevolution and, 51
Hawaii, 6, 7
Hays, James D., 39
Hemoglobin, 25, 26-27, 28
Hennig, Willi, 57
Heterotrophs, 14
"Hopeful monsters", 4, 5
Horses, evolution of, 2, 57
Hsu, Kenneth, 37, 38, 40
Humans
  arterial branching in, 9
  brain size, increase of, 5
  coevolution and, 51
  relationship to other primates, 24, 25, 26, 28-30
Hummingbirds, coevolution and, 48-49
Hutton, James, 33
Huxley, Thomas Henry, 53

Immunology, 26
Insects, fossils of, 32-33
Iridium, 35, 36, 37, 38, 39, 40

Janzen, Daniel H., 42, 45, 49-50, 51

Kiwi, 53, 55, 56, 57
Kosmoceras, evolution of, 3-4

LaBrecque, John, 37-38, 40
Lande, Russell, 4, 9
Langley, Charles H., 30-31
Life on earth, origin of, 11-20
Lyell, Charles, 33-34

McLean, Dewey M., 38-39
McNaughton, Sam J., 50-51
Macroevolution, 3, 5, 9
Macromutations, 5
Magnetic stratigraphy, 35
Mammals, evolution of, 5
Mass extinction
  Arctic Ocean spillover theory, 38
  as indicator of geological periods, 34
  asteroid theory, 35, 37, 38-40
  catastrophist theory, 32-33, 34-40
  comet theory, 39, 40
  estrogen theory, 38
  gradualist theory, 32, 33, 40
  greenhouse effect theory, 38-39
  magnetic reversal theory, 39
  of amphibeans, 34
  of Cambrian period, 34
  of Cretaceous-Tertiary period, 34-40
  of dinosaurs, 34-40
  of mammals, 38-39
  of Permian-Triassic period, 34-35, 37, 40
  of Pleistocene period, 38-39
  of reptiles, 34
  of Tertiary period, 39
  of Triassic period, 34
  of trilobites, 34
  ozonosphere theory, 38
  plate tectonics theory, 40
  supernova theory, 36
  tektite theory, 39
Mendel, Gregor, 2, 23, 42
Metazoans, 14, 20
Mice, mutations in, 3
Michel, Helen, 35-36
Microbiotas, 13
Microevolution, 3, 5, 9
Mimicry, 43, 44-45, 46-47
Modern synthesis. See Neo-Darwinism.
Molecular evolution, 4-5
  and ratites, 53, 54-56, 57
  as compared to comparative morphology, 23
  methodology of, 26
  to determine pace of evolution, 24-31
  to determine relationships of species, 23-24, 26
Mollusks, structure of, 3
Monarch butterflies, coevolution and, 43
Morphology, 2-4, 8, 9. See also Comparative morphology.
Mutation, at molecular level, 24, 55-56
Mutualism, 43, 49

Natural selection, 2, 5, 9, 30, 31
Neo-Darwinism, 2-4, 5, 7-8, 9
Neutralists, 30, 31
Neutron activation analysis, 35
Nucleotide replacement, 27, 30, 31
Nuttall, G.H.F., 23, 24

Occam's Razor, 24
Ontogenetic transformation, 57
Origin of life on earth, 11-20
Ostrich, 52, 53, 55, 56, 57
Outgroup comparison, 57

Parsimony criteria, 57
Passionflower butterflies, coevolution and, 45-47
Pauling, Linus, 24

Phylogenic trees, 21, 22, 23, 53, 54, 55, 57
Plants, speciation of, 5-6
Plate tectonics, 53-54
Pokegama formation, 18
Pollen transport, 42, 48-49
Pongola formation, 13
Population genetics, 7
Positive feedback, 42
Prager, Ellen, 56, 57
Precambrian period
  as percentage of earth's history, 13
  fossils of, 12, 13, 14-15, 16-20
Predation, 43-48, 51
Primates, evolution of, 24, 25, 26, 28-30
Procaryotes, 14, 16-17, 18, 19
Proteins, 23-24, 25, 26-27, 28, 30, 31
Protozoans, 14
Punctuated equilibrium, 3-4, 5, 7, 8

Radioisotope dating, 16, 24
Ratites
  ancestry of, 53, 54
  molecular evolution and, 53, 54-56, 57
  morphology of, 53, 55, 56-57
  phylogeny of, 53, 54, 56-57
Raup, David, 32-33, 34, 35, 40
Raven, Peter H., 42
Rhea, 52, 53, 55, 56, 57
Russell, Dale, 36, 37

Salamanders, 8, 9
Sarich, Vincent, 23, 25-26, 28, 30, 31
Schopf, J. William, 12, 13, 14, 15-19, 20, 32
Schopf, Thomas J.M., 3-4, 40
Seed transport, 43, 49
Selectionists, 30
Sibley, Charles G., 54, 56, 57
Sibling species, 8
Simpson, George Gaylord, 2, 5, 16
Speciation, 5-7
Stanley, Steven M., 20
Stebbins, G. Ledyard, 5, 8
Steep Rock Lake Formation, 18
Stromatolites, 13, 14, 15, 16, 17, 18
Survival of the fittest, 42

Tapir, coevolution and, 49-50
Teleology, 45
Tinamou, 52, 55, 57
Tinamous, 53, 56, 57
Trilobites, 11-12, 14, 34
Tyler, Stanley, 14-15

Ussher, James, 33

Van Valen, Leigh, 5, 6, 31
Viceroy butterflies, coevolution and, 44-45
Visual imprinting, 46
Volterra, Vito, 42

Wake, David, 7-8, 9
Wasps, coevolution and, 42-43
Wasson, John, 40
Wildebeeste, coevolution and, 50-51
Wilson, Allan C., 5, 25-26, 28, 30, 31, 54-55, 56, 57

Zuckerkandl, Emile, 24